Gamal Elsaeed
Mohamed Balah
Mohamed Hasan

New Design for Shore Protection Structures

Gamal Elsaeed
Mohamed Balah
Mohamed Hasan

New Design for Shore Protection Structures

Numerical Study for the Effect of Coastal Structures on Shoreline Change

LAP LAMBERT Academic Publishing

Impressum / Imprint

Bibliografische Information der Deutschen Nationalbibliothek: Die Deutsche Nationalbibliothek verzeichnet diese Publikation in der Deutschen Nationalbibliografie; detaillierte bibliografische Daten sind im Internet über http://dnb.d-nb.de abrufbar.

Alle in diesem Buch genannten Marken und Produktnamen unterliegen warenzeichen-, marken- oder patentrechtlichem Schutz bzw. sind Warenzeichen oder eingetragene Warenzeichen der jeweiligen Inhaber. Die Wiedergabe von Marken, Produktnamen, Gebrauchsnamen, Handelsnamen, Warenbezeichnungen u.s.w. in diesem Werk berechtigt auch ohne besondere Kennzeichnung nicht zu der Annahme, dass solche Namen im Sinne der Warenzeichen- und Markenschutzgesetzgebung als frei zu betrachten wären und daher von jedermann benutzt werden dürften.

Bibliographic information published by the Deutsche Nationalbibliothek: The Deutsche Nationalbibliothek lists this publication in the Deutsche Nationalbibliografie; detailed bibliographic data are available in the Internet at http://dnb.d-nb.de.

Any brand names and product names mentioned in this book are subject to trademark, brand or patent protection and are trademarks or registered trademarks of their respective holders. The use of brand names, product names, common names, trade names, product descriptions etc. even without a particular marking in this work is in no way to be construed to mean that such names may be regarded as unrestricted in respect of trademark and brand protection legislation and could thus be used by anyone.

Coverbild / Cover image: www.ingimage.com

Verlag / Publisher:
LAP LAMBERT Academic Publishing
ist ein Imprint der / is a trademark of
OmniScriptum GmbH & Co. KG
Heinrich-Böcking-Str. 6-8, 66121 Saarbrücken, Deutschland / Germany
Email: info@lap-publishing.com

Herstellung: siehe letzte Seite /
Printed at: see last page
ISBN: 978-3-659-36274-3

Copyright © 2015 OmniScriptum GmbH & Co. KG
Alle Rechte vorbehalten. / All rights reserved. Saarbrücken 2015

NEW DESIGN FOR SHORE PROTECTION STRUCTURES

By

Prof. Dr. Gamal H. Elsaeed

Prof. Dr. Mohamed I. Balah

Eng. Mohamed A. Hasan

TABLE OF CONTENTS

LIST OF TABLES

LIST OF FIGURES

LIST OF SYMBOLS

A	Dimensionless parameter
B	Beach berm height above still water level
c_x, c_y, c_θ	Propagation velocities in the energy balance equation
C_{gb}	Breaking wave group celerity
C	Group celerity
C_g	Bottom friction coefficient
D_B	Average berm height
D_C	Depth of closure
D_{LT}	Limiting depth of longshore sediment transport
D_{50}	Median grain size diameter
D_{eff}	Effective diameter of the rubble mound
d	Structure depth
E	Directional spectrum
f	Wave frequency
g	Gravitational acceleration
h	Water depth
H_0	Deep water significant wave height
H_b	Breaking wave height
H_i	Incident wave height at the tip of the breakwater
H_{rms}	Root-mean-square wave height
H_s	Significant wave height
$H_{s,12}$	Non-breaking significant wave height, that is exceeded 12 hr./year
H_{brms}	Root mean square breaker height

H_{sb}	Significant wave height at breaking
H_{sm}	Mean of the annual distribution of significant wave height
k	Wave number
K	Dimensionless empirical proportionality coefficient
K_1, K_2	Empirical coefficients in sediment transport equation
K_t	Coefficient of transmission
L_g	Groin length
L	Wave length
L_o	Deep water wave length
L_b	Breaking wave length
N	Wave-action spectrum
P	In-place porosity
Q	Longshore sediment transport rate
R_s	Stability parameter
R_c	Breakwater crest freeboard relative to the still water depth
S	Groin gap Width
S_{op}	The local wave steepness
q	Sources/sinks along the coast
t	Time
T	Wave period
T_P	Peak wave period
T_s	Significant wave period
u_{mb}	Orbital velocity
u_α	Horizontal velocity
u_f	Volume flux density
x	Alongshore distance
W	Dimensionless parameter

w_s	Fall velocity
y	Offshore distance
z	Normal direction to x-y plane
Z_α	Arbitrary elevation
α_0	Deep water wave approach angle
α_b	Breaking wave angle
α_{eb}	Efficient breaking wave angle
α_i	Incident wave angle at the tip of the structure
α_v	Incident breaking wave angle at the groin
β	Bottom slope from the shoreline to the depth of active longshore transport
ε	Transport coefficient
ζ_b	Surf similarity
ζ	Breaker parameter
η	Water-surface elevation
θ	Angle of wave approach
ρ	Water density
ρ_s	Sediment density
	Standard deviation of the annual wave height
σ	Intrinsic frequency
Φ	Velocity potential
Δt	Time increment
Δx	Alongshore grid spacing
Δy	Offshore grid spacing
Δz	Grid spacing along depth

LIST OF ABBREVIATION

1-D	One-Dimensional
2-D	Two-Dimensional
3-D	Three-Dimensional
ACD	Admiralty Chart Datum
CERC	Coastal Engineering Research Center
CoRI	Coastal Research Institute
CMS	Coastal Modeling System
EEAA	Egyptian Environmental Affairs Agency
ERDC	Engineer Research and Development Center
FHWA	U.S. Federal Highway Administration
GENESIS	Generalized Model for Simulating Shoreline Change
HRI	Hydraulics Research Institute
LCS	Low Crested Structures
MWRI	Ministry of Water Resources & Irrigation
MS	Meteorological Station
MSL	Mean Sea Level
SLR	Sea Level Rise
SMS	Surface-water Modeling System
SPA	Shore Protection Authority

ABSTRACT

The present study has been recommended the use of perched beach as a possible alternative for safe swimming conditions. Numerical simulation has been implemented by using the Surface Water Modeling System (SMS-10.1). The numerical model was applied to investigate various configurations of the perched beach including submergence ratio of the breakwater, groin with/without gap, the gap width/location and emerged/submerged groin. These configurations have been compared from the point of view of wave height, currents velocities, flushing rates and shoreline changes to develop general guidelines for the design of similar constructions.

An actual scale model of a perched beach designed and constructed to provide a safe swimming conditions. The project area is constructed along the North-West coast of Alexandria in Egypt. The latter area has long been suffered from rip currents as large as 0.7m/sec and limited safe swimming strip of less than 40m during the prevailing wave conditions in the summer season. Six alternatives of perched beaches have been studied using SMS numerical model and adopting the actual wave rose of El-Dekhila port and bathymetric survey of the project area. Field measurements have been carried out before construction of the perched beach, and it is considered as the baseline condition. The model has been calibrated and validated against the collected and measured field data. The alternatives have been simulated, the results analyzed and compared to evaluate the pilot perched beach project. Numerical results and initial field observations have shown that the constructed perched beach could be a reliable solution for protecting swimmers along the North-West coast of Alexandria in Egypt. The impacts on flushing and water quality seem to be acceptable especially with the aid of openings in the groins. The impacts of groins/breakwater on shoreline changes have been predicted and they are tolerable at reasonable annual costs.

Chapter 1
Introduction

1.1. General

Coastal zones in many countries are of major concern by the virtue of being multi-functional regions. Their use as harbors, fisheries, recreational areas, source of minerals, water supply and excess water disposal gives them high economic value. Also, the risk of low land flooding and excessive salt-water intrusion in the groundwater due to sea level rise, induced by the greenhouse effect, has become an important topic in much research. The coastal engineering field investigates the physical processes within the coastal zone and seeks a balance between various beneficiaries of it. The use of coastal structures is the tool in many cases and, hence, the impact of their construction becomes of great interest. The impact of marine structures on shoreline changes and water quality under various waves, current and site conditions is of great interest to engineers and scientists.

The Northern coastline of Egypt along the Mediterranean Sea extends for approximately 1000 km from Rafah at the east to El-Sallum at the west, as shown in Figure 1.1. The central region of the Egyptian Nile Delta extends some 300 km from a point 30 km west of Alexandria to 30 km east of Portsaid. The coastline in this area is considered to be in a state of continuous change under the action of sea waves and currents (Frihy et al, 1994). The River Nile and its two main branches, Damietta and Rosetta, supplied large volumes of sediment through the outfall of these branches to the sea. This large rate of sediment supply used to exceed the rate of sediment loss caused by wave and current action; and hence providing natural beach protection and excess sediment supply to the Nile Delta. Construction of Old Aswan Dam and High Aswan Dam significantly reduced the rate of sediment supply along the Delta shoreline causing severe erosion (John B. Herbich et al, 1996). The erosion started shortly after the construction of Old Aswan Dam, the development of other irrigation structures along the River Nile, and the reduced discharge into the sea.

After operating Aswan High Dam in 1966, erosion along the Delta coastline accelerated considerably, resulting in the loss of several beaches. The erosion had a severe effect on the nearby agricultural lands because of salt water intrusion affecting crop roots. Many hectares of beachfront properties have been lost since 1966 due to severe erosion.

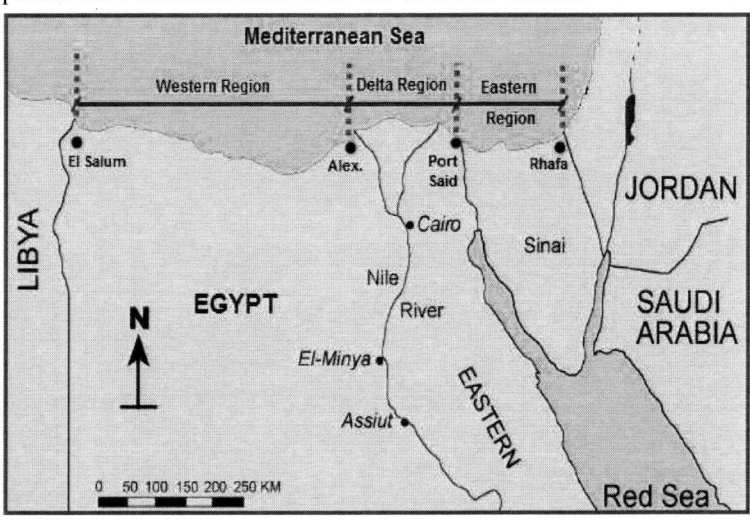

Figure 1.1: Northern coastline of Egypt along the Mediterranean Sea.

Until recently, the western region of the northern coastline outward the Nile Delta between Alexandria and El-Salum has been a natural coast with no significant human interference with the coastal system. Shoreline changes are generally small, but rip currents are serious problem to swimmers in many beaches. In the last twenty years, man has interfered with the littoral system at several locations due to the development of coastal structures in some summer resorts. With the fast increase of summer resorts, new plans are needed to provide primarily safe swimming conditions. The Mediterranean coasts of Egypt suffer from two major problems; i.e., the shoreline changes (erosion or accumulation) and dangerous swimming conditions.

The aforementioned problems have unacceptable economic impacts on summer resorts. This has initiated several researcher and engineer to search for the improvement of the swimming conditions in the area. It is also crucial to predict and monitor the impacts of any suggested coastal development project. In selecting coastal mitigation measure, it is important to evaluate and measure effectiveness and economics of the measure (s). It is also important: to know that some of the mitigation measures are able to reduce wave heights along the shoreline, or simply maintain a sacrificial beach, whereas others try to impede the longshore sand transport. In a particular situation, one mitigation measure may work better than others owing to site specific conditions. Also, some approaches can provide a net benefit to adjacent shorelines, whereas others, if misapplied, can cause severe erosion. The impacts of man-made activities/structures on the coast/near shore wave conditions are always of great interest and are obligatory to issue the approval of relevant agencies, e.g., Egyptian Environmental Affairs Agency (EEAA) and Shore Protection Authority (SPA).

1.2. Problem Definition

Strong offshore direct rip currents regularly lead to hazardous situations, and at some beaches swimming is prohibited for a considerable time of the year especially during summer storm. Also, the swimming may be prohibited at beaches where the wave height is too high. Saski et al (1975) concluded that breaker heights smaller than 0.6m and current velocities smaller than 0.2m/s are considered as comfortable swimming conditions, but it is hard to swim against a rip current of 0.5m/s and breaker height greater than 2.0m even for good swimmers.

Towards the improvement of swimming conditions, it is required to determinate the main features and processes causing the potential danger to swimmers. Two aspects have been identified as the main causes of the unfavorable swimming conditions:

a. Locations and strength of rip/undertow/long shore currents

The presence of the rip currents along the coast is the most dangerous phenomenon for swimmers. Rip currents are seaward directed flows more or less perpendicular to the coastline. The time and location of rip currents are difficult to predict, but they tend to occur most frequently in areas with sand bar (which is a dynamic feature) and when moderate waves approach the coast normally. A swimmer entering a strong rip current may float seaward to a depth where he cannot touch the seabed. Swimmer may panic and try to swim straight back to the shore against the strong rip current.

The undertow current is an offshore flow below the wave trough level which compensates for the shoreward mass transport induced by waves above the trough-level. The strength of undertow current is determined mainly by the incoming wave energy. Though under normal conditions, undertow currents are generally weak, it may cause fear to swimmers due to the seaward force produced in combination with steep beach profile. However, it is possible to swim against an undertow current since it is concentrated in the lower parts of the water column, while a swimmer can float in the upper part.

b. Steep coastal profile

Due to the steep slope of the beach profile, waves break close to the shoreline where people tend to swim. The breaking waves make it difficult for swimmers to move freely, it may even cause them to remain in place due to seaward drifts caused by the undertow and the rip currents. As a result of the adverse effect of steep beach profile, swimmers reach relatively large water depths already within a short distance from the shoreline. This, in combination with the breaking waves, allows very narrow usable strip for most swimmers along the beach.

These conditions have resulted in a growing interest to create safe beach for swimmers. Some have built swimming pools and lagoons behind or on the beach itself without taking into consideration their harmful impacts. So, there is an essential need to construct suitable coastal structure in the pilot area in order to secure safe conditions for swimmers. There are several plans and investigations for improvement of the swimming conditions. The use of coastal structures is the tool in many cases and, hence, the impact of their construction on the coast becomes of a great interest.

In Egypt, the coastal landscape west of Alexandria is one of the most attractive recreational sites. The coastline is characterized by wide beaches consisting of white Oolitic carbonate sand. The water is clear, varying from green to blue, which is a result of the white seabed sand. The shore is generally linear with few protective configurations. However, the beach is not suitable for swimming because of the steep beach profile ranging from 1:30 to 1:50 (Nafaa and Frihy, 1993).

Some resorts used surface piercing detached breakwaters for protecting the shoreline and swimmers, e.g., Marabella and Al-Nakhil resort of Egypt. In the latter cases, accretion was developed shortly after the construction of the breakwaters and the down drift zones suffered from erosion. The erosion has been dramatically increasing causing demolishment of large parts of the down drift beaches. Moreover, floats and debris are usually trapped behind surface piercing breakwaters while eddies are evident at the end sections. It is noteworthy that the latter cases violate the environmental laws 4/94 of Egypt, but they could somewhat protect swimmers from the risk of drowning.

Due to the increasing demand for safe swimming conditions with minimum impact on the shoreline and keeping acceptable water quality, new studies have been conducted to meet these requirements using appropriate coastal structures. The Shore Protection Authority of Egypt (SPA) conducted a study in 2002 for the development of the North-West coast of Egypt and recommended the use of perched beach as a possible alternative for safe swimming conditions along Al-Arab bay zone, located from Al-Agamy at Km21 to Al-Alamin city at Km 120 (along Alexandira-Matrouh road).

The perched beach consists of two bounding groins and shore parallel breakwater at groin ends enclosing a sheltered basin (Figure 1.2).However, none of the perched beach designs have been constructed to date. Thus, the actual performance of perched beach is not well-known yet. Recently in 2011, one design has been approved by the Egyptian Environmental Affairs Agency (EEAA) at Km 38 along Alexandria-Matrouh road. The construction of this perched beach started in 2012 and was complete by 2013. Thus, the actual field effect of the perched beach can be monitored and compared with the results predicted by numerical models.

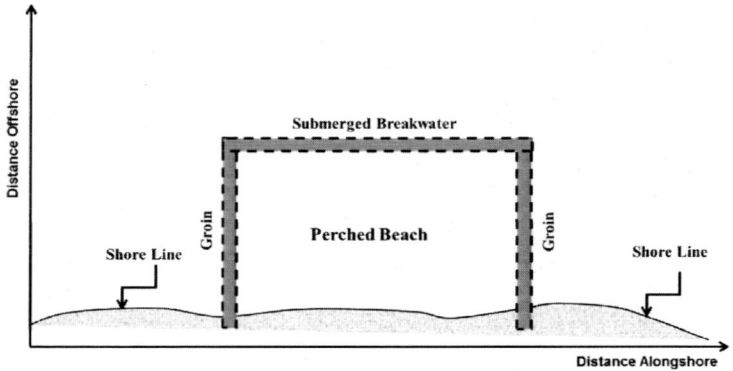

Figure 1.2: Definition sketch of the perched beach.

1.3. Objectives of the Study

The aim of the present study is summarized as follows:

- Select and apply appropriate numerical models to simulate the hydrodynamic conditions, coastal sediment transport and water quality in a proposed perched beach constructed to provide safe swimming conditions along the coast.

- Investigate various configurations of perched beaches to minimize the impacts on the shoreline and preserve acceptable water quality in the perched beach.

- Investigate the effect of various possible configurations on wave height, current velocities, flushing rates and shoreline changes within the perched beach.

- Develop guidelines for the design of perched beaches along the North-West coast of Egypt.

1.4. Methodology

In order to achieve the study goals, the following methodology has been adopted.

Numerical Simulation:

Numerical simulation has been implemented by using the Surface Water Modeling System (SMS-10.1). The numerical model was applied to investigate various configurations of the perched beach including submergence ratio of the breakwater, groin with/without gap, the gap width/location and emerged/submerged groin. These configurations have been compared from the point of view of wave height, currents velocities, flushing rates and shoreline changes to develop general guidelines for the design of similar constructions.

Based on all the collected data and information, the use of perched beach as a possible alternative for safe swimming conditions along Al-Arab bay zone in Egypt was applied. The model was calibrated and validated against the collected field data. This includes field measurements of the shorelines before/after construction, as shown in Figure 1.3.

Field Data:

Measure and monitor the impacts of the actual scale perched beach constructed at station 38Km along Alexandria-Matrouh coastal road. Field measurements have been carried out in May 2009 before the structure construction, and it is considered as the baseline condition. Also, other field measurements have been carried out during and after construction to monitor the shore line changes.

Another data have been collected about the study area from available literature. These data have been quoted from the available literature such as scientific publication by members from WL|Delft Hydraulics, the Coastal Research Institute (CoRI) and the Hydraulics Research Institute (HRI) (e.g., HRI, 2003).

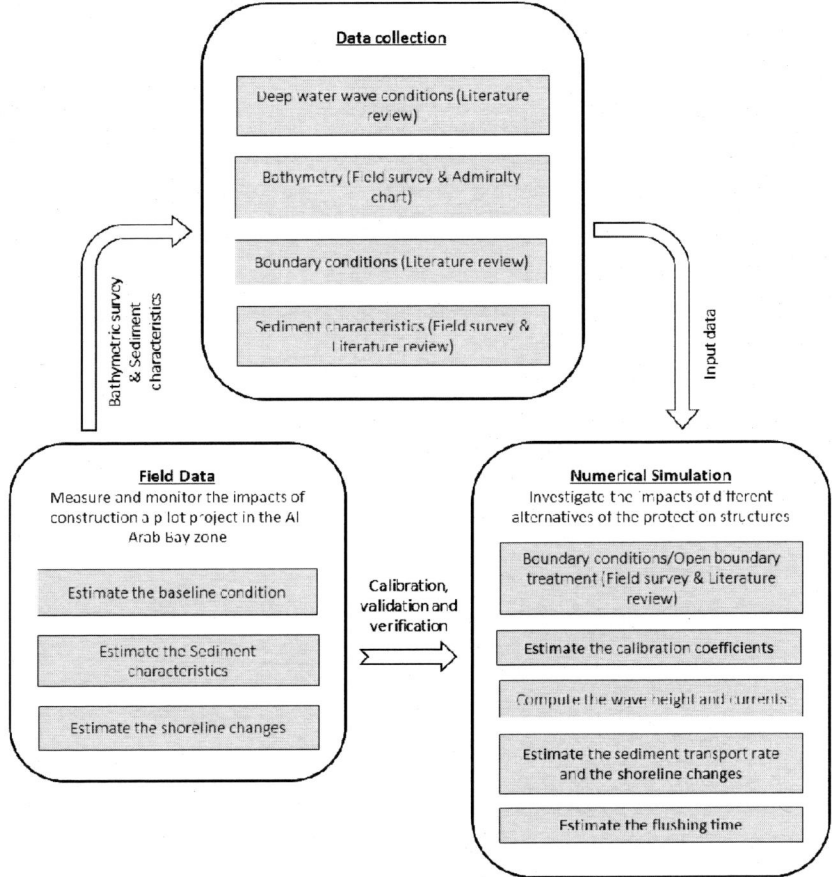

Figure 1.3: A diagram for study methodology.

1.5. Thesis Structure

Various configurations of the perched beach have been described, simulated and the results analyzed to provide general guidelines for the design of a perched beach in order to develop safe swimming areas with minimum impact on the shoreline changes and keeping acceptable water quality. Application has been made to a pilot perched beach project at station

38Km along Alexandria-Matrouh coastal road.

Chapter (2) provides a review of literature and an overview of coastal models.

Chapter (3) describes the component of numerical model (SMS-10.1), which includes wave module (CMS-Wave and BOUSS-2D), and shoreline module (GENESIS). Also, the water quality module (RMA) is described. Theoretical assumptions that backbones the model are discussed. Also, the model structure and numerical modeling such as grid system, boundary conditions and stability criterion are illustrated.

Chapter (4), Numerical models were applied to investigate various configurations of the perched beach including submergence ratio of the breakwater, groin with/without gap, the gap width/location and emerged/submerged groins. These configurations have been compared from the point of view of wave height, currents velocities, flushing rates and shoreline changes to develop general guidelines for the design of similar constructions.

Chapter (5) describes the study area and the data collection technique. The bathymetric survey, bed material distribution, sediment transports, coast dynamics, general climate conditions and prevailing wave conditions were presented. The model has been calibrated to estimate the empirical coefficients and validated against the results acquired from collected data and field measurements.

Chapter (6), application has been made to an actual scale perched beach approved in 2009 by EEAA and SPA. The perched beach was completed by 2013 and its data have been employed in the current work. Additional investigations have been conducted to extend the results and lessons learned from the existing design. Several alternatives have been examined and compared to develop general guidelines for similar beaches along the North-West coast of Alexandria. Special attention has been given to alleviate the possible adverse impact of the protection structures. These impacts include excessive shoreline erosion and possible deterioration of water quality.

In Chapter (7), a summary and conclusions of the study are presented.

Chapter 2
Literature Review

2.1. General

The planning and use of these areas without being aware of existing natural processes are often fatal. These processes are highly sophisticated dynamic events ranging from micro scale physical phenomena, such as the movement of a particular sand grain, to macro scale phenomena such as the influence of the global mean sea level rise on beach change (Hanson, 1987). Protection from waves can be done using many structures, where these structures should be constructed from durable materials such as rock or concrete, and should be designed to withstand the force of wave action. The main types of these structures include seawalls, breakwaters, revetments, groins, and jetties. Each type of these structures has its own feature and its own purpose such as seawalls built at the water's edge and its purpose to bear the full attach of the wave action. Revetment structures composed of a riprap are powerful in reducing the wave energy. Revetment surface should be irregular to offer protection from wave run-up (Revetment limits access to the beach). Groins are sediment traps at right angle from the shore and catch sediment carried by longshore drift on their up current side. Breakwaters used to attach the waves and produce calmer water shoreward of it. Breakwaters may be connected to the shoreline at one end or completely detached from it. Also, breakwaters may be aligned to the shoreline and connected with two groins which is known as a perched beach. All these structures impact littoral long shore transport causing beach build up on their up-drift sides and erosion down drift.

So, sediment transport mechanisms, wave kinematics and interactions between waves and coastal structures and bottom to epigraphy are essential investigations provide decision-makers with sufficient information about the coastal system. It helps them understand and predict the potential impact of coastal structure in the future. To investigate coastal processes and influences of human activities in coastal areas, numerous numerical models have been developed over the past decades. In this chapter an overview on perched beach and wave/sediment models are presented.

2.2. Review of Perched Beaches

The current study concentrates on the use of perched beach. This scheme consists of a submerged breakwater enclosing a sheltered basin. The top level of the offshore part of the submerged breakwater is close to the sea water surface level to ensure water flow from open sea to the basin and not to obstruct sea view; i.e. minimal visual impact. Basins enclosed by submerged breakwaters are existing naturally; i.e. rocky submerged shoal enclosing a shelter basin such as in Stanley beach at Alexandria named "Al Bahr al Sagir" and artificially or semi-artificially such as in Sela beach at Bat-Yam, as illustrated in Figure 2.1.

Figure 2.1: Sela beach development scheme at bat-yam, constructed since 1969.

The enclosed basin at Sela beach is 400 meters alongshore by 175 meters perpendicular to the shoreline. The submerged breakwater at Sela beach was built in 1969 at water depths of approximately 2.5 to 3.0 meters with a nominal top level equal to mean sea water surface level. The littoral drift bypass around the breakwater resulting in negligible impact on neighboring shoreline evolution due to breakwater construction.

Periodic water quality tests had been performed showing that water quality inside the enclosed basin is similar to open sea due to the outflow of water mainly through the rubble mound breakwater as still water level is higher inside than outside the basin. Wave height inside the basin enclosed by the submerged breakwater were significantly reduced compared to the open sea. Wave heights are about 0.5 meter inside the basin compared to 1.5 meters outside the basin. During winter storms wave heights in the basin may reach up to 1.0 to 1.5 meter and currents of significant velocities were reported by good swimmers (Tauman J., 1976).

Prior to the construction of Sela beach breakwater the nearshore bar was sounded and found to have a crest elevation at about 1.0m below sea water surface level quasi-parallel to the shoreline as illustrated in Figure 2.2. Wave breaking occurs on the sea side or on the top (crest) of the sand bar and almost the entire width of the foreshore is covered by surf. The area between the coastline and the inner (onshore) side of the bar acts as a velocity field in which flow takes place in all directions (Figure 2.2). Along the nearshore bar there are breaks (channels) or rip passes through which rip currents are forced out to the sea. On top of the bar the water seemed to be air entrained absorbing currents from both sides. The bar configuration, rip passes, currents velocities and directions are changing under different waves conditions. This flow field, the rhythmic bathymetry and the difficulty of returning to the shore due to rip currents are the reasons of drowning accidents; i.e. conditions very similar to the northwestern Egyptian Mediterranean coast.

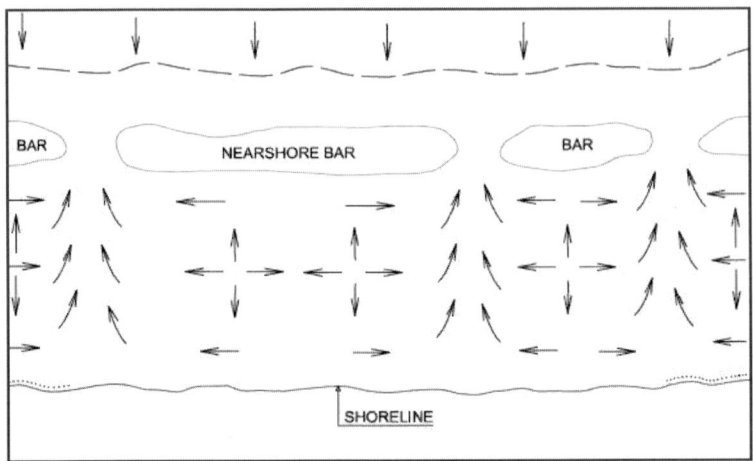

Figure 2.2: Schematic diagram of nearshore bar and nearshore current.

The perched beach scheme was tested by Delft Hydraulics, The Netherlands. In 2003, they studied several possible alternatives along the Arab Bay of Egypt to develop safe swimming conditions. They focused on the following options:

Option 1: submerged breakwater with high (emerged) end-groins,

Option 2: submerged breakwater with low (submerged) end-groins.

They concluded that option 1 is more suitable for use in the study area. However, in the present study various configurations of the perched beach have been studied and evaluated. These include breakwater height, groin with/without gap and the gap width/location. The evaluation has been carried out with emphasis on swimming conditions as well as coastal impact. Other aspects, such as, aesthetics and flushing rates, have been considered and general guidelines for the design of perched beaches have been developed.

2.3. Review on Wave Models

Wave models could be classified into:1) phase-resolving models (deterministic), and 2) phase-averaging models (stochastic). In the first type, the fluctuating instantaneous sea surface is directly resolved, i.e. the surface is covered with a grid which must be sufficiently fine relative to the wave length. The individual waves in a spectrum are resolved as per their phase. The phase-resolving models are used when average properties of waves change rapidly and when there is rapid variations in depth and shoreline. Among the phase-resolving wave models; Boundary integral models, Mild Slope Equation model, Shallow Water Equation model, and Boussinesq Equation models. The mild slope equation is weekly nonlinear and is applicable to mild bed slope.

The Boussinesq model consists of one continuity equation and one momentum equation. It was derived by (Peregrine, 1966) and it was restricted to shallow water with weak nonlinearity. Various attempts have been made to improve the applicability of Boussinesq equations by improving the frequency dispersion characteristics, Madsen et al.(2006), and by including the vortex effect. The BOUSS-2D model is one of the famous models following the concept of phase-resolving. BOUSS-2D model is based on Boussinesq-type equations derived by Nwogu (1993, 1996). The BOUSS-2D model is capable to describe the shoaling, diffraction, and refraction at the intra-wave scale (fine resolution in space and time). Nonlinearity is accounted for and the computation of the current could be done implicitly.

In the second type, wave models, are commonly based on Energy Balance equation and Parabolic equations. The phase-averaged wave energy is computed. These models consider evolution of a directional and frequency spectrum. These models predict averaged or integral properties like significant wave height, average wave period, and wave energy etc. These models are used when average properties change slowly over few wave lengths. The phase average models have the advantages of being quite stable and having less computational time (coarse resolution). It is highly dispersive and the generation, dissipation and wave-wave interaction could be included. Drawbacks of these models in coastal water include the use of linear wave theory for wave propagation and explicit computation for the current. Also, for detailed investigation of the beach morphology in the nearshore zone, it is better to use the phase resolving models.

Horikawa and Kuo (1966) studied the linear wave transformation inside the surf zone analytically based on the concept of the phase average model. Dally and Dean (1985) investigated the intuitive expression for the spatial change in energy flux associated with wave breaking in the surf zone. Tajima and Madsen (2006)incorporated the nonlinear effects by establishing a correspondence between linear and non-linear wave characteristics.

Several studies followed the same concept carried out. Three generations were developed until now based on this concept. The WAM, SWAN, STWAVE and CMS-Wave are among the most famous models following the concept of phase average. Originally they don't count for diffraction. Recently diffraction is included by different ways based on approximation from Mild Slope Equation or approximation based on Parabolic Equation. (Mase, 2001)developed a random wave transformation model in which diffraction effects were included. The diffraction term was introduced by utilizing the parabolic approximation of the wave equation. The numerical scheme was found to be stable (the first order upwind finite difference). Mase et al. (2005) improved the model by using Quadratic upstream interpolation for convective kinematics in the discretization to reduce numerical diffusion, in addition to taking into account reflection in the model.

The main advantages of using the Mase's model (CMS-Wave) are the ability to reproduce wave transformation over complicated bathymetry and setting up the input and output for many different wave spectra.

In the present study, the wave models BOUSS-2D was coupled with CMS-Wave to simulate the wave transformation. CMS-Wave model calculates the change in the local wave energy occurring during transformation of waves from deep water to the shallow water project site, while the actual wave height in certain areas, particularly near structures, can be predicted more accurately by BOUSS-2D model.

2.4. Overview on Coastal Sediment Models

Depending on the coastal processes and scope of the study, morphological models expand from simple 1-D to sophisticated 3-D models. The available numerical models can be classified under three main categories; shoreline change models, beach profile models, and 3-D models (Dabees, 2000).The uses of numerical models for different spatial and temporal scales are given in Figure 2.3(Hanson and Kraus, 2011). The medium-term beach models are classified with respect to their longshore extent, time range and cross-shore extent, as shown in Figure 2.3.

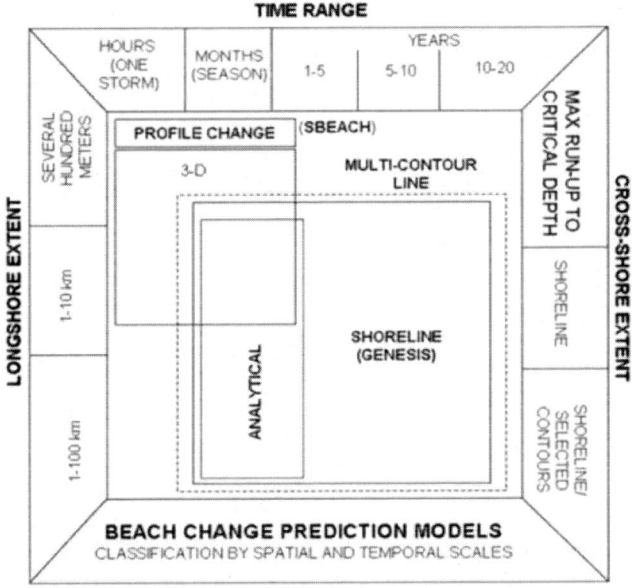

Figure 2.3: Classification of beach change models by spatial and temporal scales (modified from Hanson and Kraus, 2011)

The fundamentals of shoreline change models were first established by Pelnard-Considere (1956), who set down the basic assumptions of the "one-line" theory, derived a mathematical model, and compared the solution of shoreline change at a groin with laboratory experiments.

Bakker (1968) extended the concept to account for possible on–offshore transport and formulated a two-line schematization of the profile. Additional contributions to such models have been produced by Fleming and Hunt (1976) for the bathymetry modification as a change in depths at a set of schematized grid points and by LeMéhauté and Soldate (1978) for the inclusion of wave refraction and diffraction.

One-dimensional shoreline evolution models have demonstrated their predictive capabilities in numerous projects (Hanson, 1987).Hanson (1987) developed GENESIS (GENEralize model for SImulating Shoreline changes), which contributes to the one line model with new features on how to dealwith structures in a shoreline change model which is applicable under several boundaries and constraints.

In the present study, the GENESIS model is chosen for the simulation of the shoreline changes and evaluate the sediment rate. A one-dimensional model approach is preferred due to the facts that this type of models have wide ranged applicability in temporal and spatial scales, requires less detailed input data and computer time and may give both qualitatively and quantitatively acceptable results which may be used for both engineering and scientific purposes.

2.5. Review on Flushing Rates

Flushing is the physical exchange of water mass between one water body and another by the advection and diffusion processes. The retention time is a commonly used measure for assessment of the potential for water quality conditions. Unfortunately there is no stringent definition of retention time in the general case of advection-dispersion processes. Prandle (1984) showed that under certain assumptions the flushing time is equivalent to the time taken for the mass level to fall to 37% of the initial level. Goshow et al. (2008), proposed a relation between the flushing time and the surface area of the water body (Figure 2.4). The flushing time in this figure is the time for the concentration to drop to 10% of its initial value. This relation will be used to estimate the acceptable flushing time in the present study.

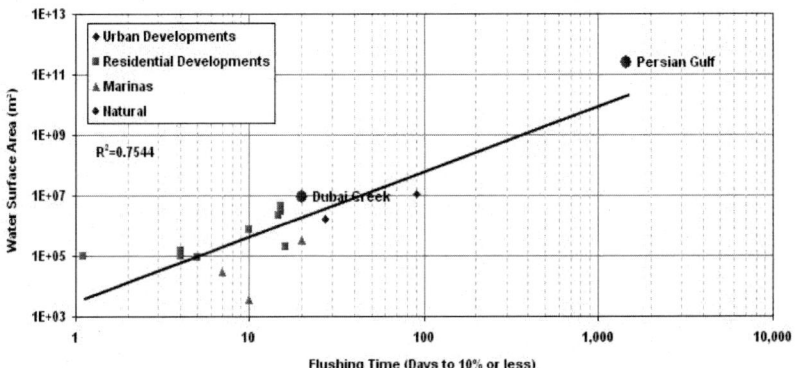

Fig. 2.4: Relation between water surface area and the flushing time for water bodies (Goshow et al., 2008).

By assuming that the flood flow entering the semi-enclosed inlet behaves like a jet, while the flow leaving the bay during ebb tide is a sink-type of flow (Stommel and Farmer 1952; Fischer et al. 1979), an analytical model of the tidal exchange characteristics between the inlet and its exterior can be obtained. However, the simple method is limited to inlets with one entrance, and for regular geometries. By the nature of the tidal prism method, density stratification and shear dispersion effects on the flushing time cannot be accounted for.

With the advance of numerical hydrodynamic and mass transport models, tidal flushing times have also been determined via numerical models (e.g. Prandle 1984; Choi et al.1989; Signell and Butman 1992; Luff and Pohlmann 1995). The basic idea is that a mass of hypothetical conservative tracer is instantaneously introduced into a region of interest. A unit tracer concentration is prescribed initially inside that region, and the subsequent a dvective dispersion of this mass is then obtained by solving the mass transport equation numerically and the time-variation of the tracer mass inside the region is tracked.

The effect of tide and wind on the physics of flushing and the flushing time in Boston Harbour was investigated using a 2D numerical model by Signell and Butman (1992). Oliveria and Baptista (1996) proposed the use of a cumulative histogram of flushing times (for different sub-areas within the region) instead of a single flushing time. Monsen et al. (2002) used numerical tracer experiments to determine the flushing time and residence time for a shallow tidal lake.

In both of the above studies, the particle tracking method is used to simulate the tracer mass transport and the effect of turbulent diffusion is not included. The effect of density stratification has not been studied in all of the previous numerical studies. While, the effect of tide, wind, density stratification and turbulent diffusion on the physics of flushing are included in the RMA model (King et al., 1995).

The RMA model is a finite element modeling system for two- and three-dimensional free surface flow. The model has two modules; a hydrodynamic module (RMA-10) and a Water Quality module (RMA-11). RMA-10 is designed for prediction of flow and depth (plus water temperature, salinity and/or suspended sediment if needed) for stratified flow systems where there are significant variations of current over the depth. This model is fully three-dimensional. It can also be used for prediction of current directions and magnitudes and water depth in system that can be depth averaged (2-D horizontal) or area averaged (1-D horizontal). Typical systems are fully mixed estuaries and coastal systems, rivers and floodplains.RMA-11 is a multi-component water quality model designed to take the currents/depths from RMA-10 and apply an advection diffusion equation to predict the transport of constituents. Simultaneously settling and chemical kinetics is introduced to predict transformation of the various constituents. In this way the various nutrient cycles (Oxygen, Phosphorous, and Nitrogen) and algae may be simulated.

This model has been designed so that the constituents and reactions could be easily changed to suit individual locations and applications. This model is capable of one/two/three-dimensional approximations in any combination. It also has a bed model capable of tracking the evolution cohesive or non-cohesive sediments. Based on the aforementioned reasons, the RMA Modeling Suite is chosen to investigate the flushing rates in the present study.

Chapter 3
The Numerical Models
3.1. General

The Surface-water Modeling System (SMS) is a comprehensive environment for one- and two-dimensional models dealing with surface water applications. Hydrodynamic models include CMS-Flow, TABS (RMA2, RMA4), FESWMS, ADCIRC, and TUFLOW. The hydrodynamic models cover a range of applications including river flow analysis, rural and urban flooding, estuary and inlet modeling, and modeling of large coastal domains. Additional functionalities include advection/diffusion (RMA) and sediment transport (FESWMS). Wave models in SMS include CMS-Wave, STWAVE, BOUSS-2D, and CGWAVE and include both spectral and wave transformational models. SMS originally developed by Brigham Young University (1985) in cooperation with the U.S. Army Corps of Engineers, Engineer Research and Development Center (ERDC), and the U.S. Federal Highway Administration (FHWA).

In the present study, the simulation consists of three parts, wave models (CMS-Wave andBOUSS-2D), shoreline/sediment transport model (GENESIS) and water quality model (RMA).

3.2. Wave Models

Numerical wave models BOUSS-2D and CMS-Wave may be used together (coupled) to evaluate potential alternatives of coastal planning for various coastal conditions. The theoretical background and user manuals for BOUSS-2D are available in CMS technical reports and CHETNs (Demirbilek et al., 2005a, b; Nwogu and Demirbilek, 2001).

CMS-Wave (Lin et al., 2008; Lin et al., 2011; Demirbilek et al., 2007) is part of the Coastal Modeling System (CMS) for simulating combined waves, currents, sediment transport, and morphology change at coastal inlets, estuaries, and river mouths (Demirbilek and Rosati, 2011). These models can simulate wave processes in navigation channels, erosion problems at coastal inlets, and aid in design and operation of harbors, port expansion, and infrastructure modifications.

BOUSS-2D is a two dimensional (2-D) phase-resolving wave model that employs a time-domain solution of fully nonlinear Boussinesq-type equations for waves propagating in water of variable depth. CMS-Wave, on the other hand, is a 2-D phase-averaged, steady-state spectral wave transformation model based on the wave-action balance equation. For harbor applications, in addition to modeling wave shoaling, bottom friction, wave breaking and dissipation processes, these models represent combined wave diffraction-refraction, full/partial reflection and transmission, nonlinear wave-wave interactions, infra-gravity waves, wave run up and overtopping of structures, and wave current interactions. While BOUSS-2D calculates both waves and wave-induced currents simultaneously, CMS-Wave calculates wave-induced currents by coupling with the CMS-Flow (Demirbilek and Rosati, 2011), a hydrodynamic and sediment transport model. Both wave model shave been verified and validated with laboratory and field data and have been applied extensively in many Corps projects (Demirbilek and Rosati, 2011; Lin et al., 2011).

In the present study, BOUSS-2D model was coupled to CMS-Wave model. CMS-Wave model is a spectral model that calculates the change in the local wave energy (but not the wave phase information) occurring during transformation of waves from deepwater to the shallow water project site. Significant wave heights calculated with CMS-Wave is a meaningful measure of the change in local wave energy level, while the actual wave height in certain areas, particularly near structures, can be predicted more accurately by BOUSS-2D.

However, because BOUSS-2D generally requires comparatively longer computational times and smaller spatial domains, CMS-Wave is applied regularly to obtain estimates of wave parameters for project planning and evaluation studies. The estimates for final design in shallow water may be verified with BOUSS-2D, which is an advanced wave model capable of representing complex physics of nonlinear wave processes nearshore. For large spatial coastal applications, CMS-Wave could be applied in a nested mode using two grids: a coarse parent grid in the open coast area that provides input wave conditions to a more refined child grid covering the vicinity of coastal structures. The child grid may also be used as a BOUSS-2D grid. The saved wave parameters or two-dimensional wave spectra from the parent grid at the boundary of child grid may also be used as input to BOUSS-2D for a detailed investigation of waves in planning, navigation and flood damage reduction studies. In this modeling approach, CMS-Wave performs a dual function:

▪ It uses a regional scale larger and coarser grid to transform deepwater wave conditions to the project site and provides boundary conditions forBOUSS-2D.

▪ Uses the CMS-Wave child grid over a smaller local domain that is highly refined to calculate wave parameters nearshore.

The offshore boundary of the BOUSS-2D grid should be placed shoreward of the location where water depth begins to influence wave propagation (e.g., depth approximately less than half the wavelength). A proper child grid of CMS-Wave may also be used as BOUSS-2D grid, with minor adjustments.

3.2.1. CMS-Wave Model

The CMS-Wave is a spectral wave model belonging to the phase-averaged class (Lin et al., 2008; Smith et al., 2001; Booij et al., 1999). It is commonly based on Energy Balance equation. It performs steady-state spectral transformation of directional random waves co-existing with ambient currents in the coastal zone. The model simulates half-plane and full-plane wave propagation, so that wave generation, wave reflection and bottom frictional dissipation of multi-directional waves can be considered. To facilitate the application of CMS-Wave, the model has a user-friendly interface in the Surface-water Modeling System (SMS). CMS-Wave calculates the spectral wave transformation based on the wave-action balance equation (Lin et al., 2008):

$$\frac{\partial C_x N}{\partial x} + \frac{\partial C_y N}{\partial y} + \frac{\partial C_\theta N}{\partial \theta} =$$

$$\frac{\kappa}{2\sigma}\left[\left(CC_g \cos^2 \theta N_y\right)_y - \frac{CC_g}{2}\cos^2 \theta N_{yy}\right] + S_{in} + S_{dp} + S_{nl}$$

(3.1)

Where, $N=E/\sigma$ is the frequency and direction dependent wave-action spectrum, defined as the wave directional spectrum $E = E(x,y,\sigma,\theta)$ divided by the intrinsic frequency σ. N_y and N_{yy} denote the first and second derivatives with respect to y; x and y are the horizontal coordinates; θ is the wave direction measured counterclockwise from the x-axis; C and C_g are wave celerity and group velocity; C_x, C_y, and C_θ are the characteristic velocity with respect to x, y, and θ, respectively. κ is an empirical parameter representing the intensity of wave diffraction effect. The right-hand side terms respectively are: S_{in} is the source (e.g., wind input), S_{dp} is the sink (e.g., bottom friction, wave breaking, white capping, etc.), and S_{nl} is the nonlinear wave-wave interaction. The characteristic velocity C_x, C_y, and C_θ are defined as follows:

$$c_x = C_g \cos \theta \qquad\qquad\qquad (3.2)$$
$$c_y = C_g \sin \theta \qquad\qquad\qquad (3.3)$$
$$c_\theta = \frac{\sigma}{\sinh 2kh}\left(\frac{\partial h}{\partial x}\sin\theta - \frac{\partial h}{\partial y}\cos\theta\right) \qquad (3.4)$$

σ is the intrinsic wave frequency. The dispersion relation of waves with uniform current is expressed as:

$$\sigma^2 = gk \tanh kh \qquad\qquad\qquad (3.5)$$

The first term on the right hand side of Equation 3.1 is the wave diffraction term (κ) formulated from a parabolic approximation wave theory (Mase, 2001). In applications, the diffraction intensity parameter κ (≥ 0) needs to be calibrated and optimized for featured structures. The model omits the diffraction effect for $\kappa = 0$ and calculates the diffraction for $\kappa > 0$. In practice, the value of κ may range from 0 (no diffraction) to 4 (strong diffraction) for calculating diffraction effects. A constant value of κ

= 2.5 has been used by Mase et al. (2001, 2005a, 2005b) to simulate wave diffraction for narrow and wide gap breakwater applications. Lin et al. (2008) and Demirbilek et al. (2009) demonstrated that value of $\kappa = 4$ is appropriate for semi-inifinite long breakwaters and also in narrow gaps (inlets) with openings equal or less than one wave length. For wider gaps with the opening greater than one wave length, $\kappa = 3$ is recommended. The exact value of κ in an application is dependent on the structure's geometry, local bathymetry and incident wave conditions, and may need to be fine-tuned with data.

This brief description of the governing equations provides the general mathematical basis for the CMS-Wave model. The detailed description of the governing equations, parameter definitions, and numerical implementation are presented in the CMS-Wave technical report (Lin et al., 2008), including a number of examples of practical applications.

3.2.2. BOUSS-2D Model

BOUSS-2Dmodel is based on Boussinesq-type equations derived by Nwogu (1993, 1996).The equations are depth-integrated equations for the conservation of mass and momentum for nonlinear waves propagating in shallow and intermediate water depths. They can be considered to be a perturbation from the shallow-water equations, which are often used to simulate tidal flows in coastal regions. For short-period waves, the horizontal velocities are no longer uniform over depth and the pressure is nonhydrostatic. The vertical profile of the flow field is obtained by expanding the velocity potential, Φ, as a Taylor series about an arbitrary elevation, Z_α, in the water column. For waves with length, L, much longer than the water depth, h, the series is truncated at second order resulting in a quadratic variation of the velocity potential over depth:

$$\Phi(x,z,t) = \phi_\alpha + \mu^2(z_\alpha - z)[\nabla\phi_\alpha \cdot \nabla h] + \frac{\mu^2}{2}\left[(z_\alpha + h)^2 - (z + h)^2\right]\nabla^2\phi_\alpha + O(\mu^4) \tag{3.6}$$

Where $\tilde{\Phi}_\alpha = \Phi$ (x, Z_α, t), $\nabla = (\partial/\partial x, \partial/\partial y)$, and $\mu = h/L$ is a measure of frequency dispersion. The horizontal and vertical velocities are obtained from the velocity potential as:

$$u(x,z,t) \;=\; \nabla\Phi \;=\; u_\alpha + (z_\alpha - z)\left[\nabla(u_\alpha \cdot \nabla h) + (\nabla \cdot u_\alpha)\nabla h\right]$$
$$+ \;\frac{1}{2}\left[(z_\alpha + h)^2 - (z + h)^2\right]\nabla(\nabla \cdot u_\alpha) \tag{3.7}$$

$$w(x,z,t) \;=\; \frac{\partial \Phi}{\partial z} \;=\; -\left[u_\alpha \cdot \nabla h + (z + h)\nabla \cdot u_\alpha\right] \tag{3.8}$$

Where $u_x = \nabla\Phi|_{z_x}$ is the horizontal velocity at $z = z_\alpha$. Given a vertical profile for the flow field, the continuity and Euler (momentum) equations can be integrated over depth, reducing the three-dimensional problem to two dimensions. For weakly nonlinear waves with height, H, much smaller than the water depth, h, the vertically integrated equations are written in terms of the water-surface elevation $\eta(x,t)$ and velocity $u_\alpha = (x,t)$ as (Nwogu 1993):

$$\eta_t + \nabla \cdot u_f \;=\; 0 \tag{3.9}$$

$$u_{\alpha,t} + g\nabla\eta + (u_\alpha \cdot \nabla)u_\alpha + z_\alpha\left[\nabla(u_{\alpha,t} \cdot \nabla h) + (\nabla \cdot u_{\alpha,t})\nabla h\right]$$
$$+ \;\frac{1}{2}\left[(z_\alpha + h)^2 - h^2\right]\nabla(\nabla \cdot u_{\alpha,t}) \;=\; 0 \tag{3.10}$$

Where g is the gravitational acceleration and u_f is the volume flux density given by:

$$u_f \;=\; \int_{-h}^{\eta} u \, dz \;=\; (h + \eta)u_\alpha + h\left(z_\alpha + \frac{h}{2}\right)\left[\nabla(u_\alpha \cdot \nabla h) + (\nabla \cdot u_\alpha)\nabla h\right]$$
$$+ \; h\left[\frac{(z_\alpha + h)^2}{2} - \frac{h^2}{6}\right]\nabla(\nabla \cdot u_\alpha) \tag{3.11}$$

The depth-integrated equations are able to describe the propagation and transformation of irregular multidirectional waves over water of variable depth. The elevation of the velocity variable z_α, is a free parameter and is chosen to minimize the differences between the linear dispersion characteristics of the model and the exact dispersion relation for small amplitude waves. The optimal value, $z_\alpha = -0.535h$, is close to mid-depth.

For steep near-breaking waves in shallow water, the wave height becomes of the order of the water depth and the weakly nonlinear assumption made in deriving Equations 3.9 and 3.10 is no longer valid. Wei et al. (1995) derived a fully nonlinear form of the equations from the dynamic free surface boundary condition by retaining all nonlinear terms, up to the order of truncation of the dispersive terms. Nwogu (1996) derived a more compact form of the equations by expressing some of the nonlinear terms as a function of the velocity at the free surface, u_η instead of u_f. Additional changes have also been made to the equations to allow for weakly rotational flows in the horizontal plane and ensure that z_α, remains in the water column for steep waves near the shoreline and during the wave run up process. The revised form of the fully nonlinear equations can be written as:

$$\eta_t + \nabla \cdot u_f = 0 \tag{3.12}$$

$$
\begin{aligned}
&u_{\alpha,t} + g\nabla\eta + \left(u_\eta \cdot \nabla\right)u_\eta + w_\eta \nabla w_\eta + (z_\alpha - \eta)\left[\nabla\left(u_{\alpha,t} \cdot \nabla h\right) + (\nabla \cdot u_{\alpha,t})\nabla h\right] \\
&\quad + \frac{1}{2}\left[(z_\alpha + h)^2 - (h+\eta)^2\right]\nabla\left(\nabla \cdot u_{\alpha,t}\right) \\
&\quad - \left[\left(u_{\alpha,t} \cdot \nabla h\right) + (h+\eta)\nabla \cdot u_{\alpha,t}\right]\nabla\eta \\
&\quad + \left[\nabla\left(u_{\alpha,t} \cdot \nabla h\right) + (\nabla \cdot u_{\alpha,t})\nabla h + (z_\alpha + h)\nabla\left(\nabla \cdot u_\alpha\right)\right]z_{\alpha,t} = 0
\end{aligned}
\tag{3.13}
$$

Where $z\alpha$ is now a function of time and is given by $z_\alpha + h = 0.465(h + \eta)$. The volume flux density u_f is given by:

$$
\begin{aligned}
u_f &= (h+\eta)\left\{u_\alpha + \left[(z_\alpha + h) - \frac{(h+\eta)}{2}\right]\left[\nabla(u_\alpha \cdot \nabla h) + (\nabla \cdot u_\alpha)\nabla h\right]\right. \\
&\quad \left. + \left[\frac{(z_\alpha + h)^2}{2} - \frac{(h+\eta)^2}{6}\right]\nabla(\nabla \cdot u_\alpha)\right\}
\end{aligned}
\tag{3.14}
$$

The fully nonlinear equations are able to implicitly model the effects of wave current interaction. Currents can either be introduced through the boundaries or by explicitly specifying a current field, U.

3.3. GENESIS Model

GENESIS (GENEralize model for Simulating Shoreline changes) was developed to simulate long-term shoreline change on an open coast as produced by spatial and temporal differences in longshore sand transport (Hanson, 1987and 1989; Hanson and Kraus, 1989). The GENESIS model is based on the one line theory. "One-line" theory was first introduced by Pelnard-Considere (1956). The fundamental assumption of the theory is the concept of "equilibrium beach profile", which cross-shore transport effects such as storm-induced erosion and cyclical movement of shoreline position associated with seasonal changes in wave climate are assumed to cancel over along simulation period and the migration of shoreline position in time is due to longshore sediment transport only. Although equilibrium beach profile assumption is verified by numerous observations, in cases such that excessive erosion happens in front of a seawall, bottom slope changes and equilibrium profile vanishes along the seawall (Hanson and Kraus, 1986). Another important assumption is that the longshore sediment transport is limited over an active depth from berm height at the shore side to a certain depth called as "depth of closure" at the sea side (Capobianco et al., 2002).

Following the assumptions of "one-line" theory, a mathematical model for long- term shoreline evolution can be described by a conservation of mass equation (3.21) and an equation of sediment transport for a sandy beach system (3.22-3.30) or (3.23). As the principle of mass conservation applies to the system (Figure 3.1)at all times, the following differential equation is obtained,

$$\frac{\partial y}{\partial t} + \frac{1}{(D_B + D_C)} \left(\frac{\partial Q}{\partial x} - q \right) = 0 \qquad (3.21)$$

Where y is the shoreline position, x is the longshore coordinate, t is the time, Q is the longshore sand transport, q represents sand sources or losses along the coast (such as river discharges, beach nourishment or net cross-shore sand loss), D_C is the depth of closure and D_B is the berm height.

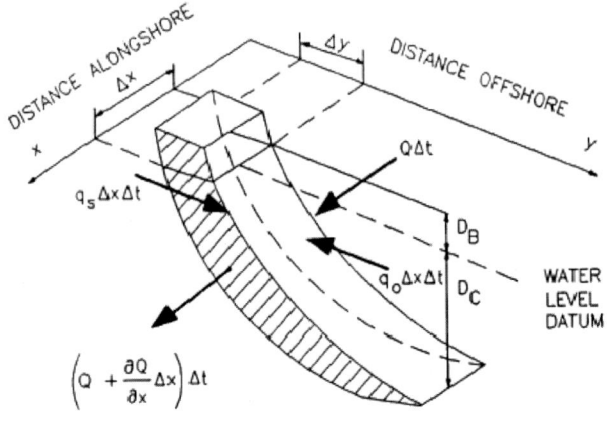

a. Cross-section view

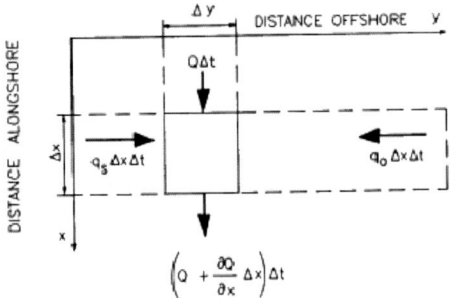

b. Plan view

Figure 3.1: Definition sketch for conservation of mass for a sandy beach system.

The closure depth could be estimated by (Hallermeier and Belvoir, 1978), (3.22) or (Hallermeier, 1981), (3.24) or (Birkemeier, 1985), (3.25)

$$D_C = 2.28H_{s,12} - \frac{68.5(H_{s,12})^2}{gT^2} \tag{3.22}$$

Where, $H_{s,12}$ is the maximum wave height of non-breaking waves that occur more than 12 hours per year(0.137 %). Hanson, (1987) used the H_b instead of $H_{s,12}$.

$$H_{s.12} = H_s + 5.6\delta \tag{3.23}$$

δ is the standard deviation of the annual wave height.

$$D_C = H_{sm}T\sqrt{\frac{g}{5000 D_{50}}} \tag{3.24}$$

Where, H_{sm} is the mean of the annual distribution of significant wave height, T_s is the corresponding period, and D_{50} is the median grain size of the beach sand in mm

$$D_C \cong (1.5 \leftrightarrow 2.0)1.75 H_s - 57.9(H_s^2 / gT_s^2) \tag{3.25}$$

Omar et al. (2005) derived several equations for the closure depth from Abu Qir up to Baltim resort based on the significant wave height and wave period and sediment size, (3.26) to (3.28).

$$D_C = 43.89 H_s - 20.9 \tag{3.26}$$

$$D_C = 47.92 H_s - 0.028 gT_s^2 - 7.6 \tag{3.27}$$

$$D_C = 27.37 H_s - 0.045 gT_s^2 - 124.08 D_{50} - 11.5 \tag{3.28}$$

Where, H_s is the mean significant wave height, T_s is the significant wave period and g is the acceleration of gravity. In the current study, D_c is taken 8m.

For the calculation of longshore sediment transport rate, Ozasa and Brampton (1980) formula has been used. In this equation, alongshore wave height gradients were considered. Such situations often occur when waves diffract around a headland or a breakwater.

$$Q = \frac{H_b^2 C_{gb}}{8(\rho_s / \rho - 1)(1 - p)w_s}\left[\frac{K_1 \sin(2\alpha_{bs})}{2} - \frac{K_2 \cos \alpha_{bs}}{\tan \beta}\frac{\partial H}{\partial x}\right]_b \tag{3.31}$$

Where, H_b is breaking wave height (m); C_{gb} is the breaking wave group celerity (m/s); α_{bs}, angle of breaking waves to the local shoreline; K_1 and K_2 are empirical coefficients, treated as calibration parameters and β is the bottom slope from the shoreline to the depth of active longshore transport.

In the current study, the sediment transport was calculated based on (Ozasa and Brompoton, 1980) to take the effect of varying breaking height due to the structure. Another advantage is that it takes into consideration the density of the sediment grains which may vary for beaches composed of coral sands, coal, etc. Further discussion on some available longshore sediment transport formulas is presented in Artagan (2006).The calibration coefficients were calibrated using the shoreline data acquired from the bathymetric survey.

- **The calibration coefficients of the longshore sediment transport formulas**

The calibration coefficients are a key factor in the longshore sediment transport formulas. However reference information at different sites are needed to calibrate these parameters, sometimes this information are not available. Different studies recommend some values for these calibration coefficients. These values are mainly estimated from field measurements in the dynamic surf zone which is non-controllable and non-repeatable and might lead to large uncertainties. Values derived from laboratory experiment have less uncertainties but the scale effect is one of the difficulties facing this approach which might lead to unreliable values. Various studies linked the K value to different features along the nearshore zone like the grain size characteristics (median grain size, falling velocity) or wave characteristics (breaking wave angle, orbital velocity, and surf similarity). Swart (1976) developed Equation (3.32) which describes K in terms of sediment size.

$$K = 1.15\log_{10}\left(\frac{0.00148}{D_{50}}\right) \qquad 0.1x10^{-3} < D_{50} < 1.0x10^{-3}m \qquad (3.32)$$

Bailard (1984) developed equation(3.33)in which K represents a function of the breaking wave angle and ratio of orbital velocity magnitude and the fall velocity.

$$K = 0.05 + 2.6\sin^2(2\alpha_b) + 0.007\frac{u_{mb}}{w_f} \tag{3.33}$$

Where in the shallow water,

$$u_{mb} = \frac{k}{2}\sqrt{gd_b} \tag{3.34}$$

The data range used for the derivation of Ballard equation is listed as follow:

$2.5 \le w_f \,(fall\ velocity) \le 20.5 cm/\sec;$

$0.2^\circ \le \alpha_b \,(breaking\ wave\ angle) \le 15^\circ;and$ (King, 2005)

$33 \le u_{mb} \,(orbital\ velocity) \le 283$

Valle et al. (1993) derived an empirical relation of K based on data from the Adra River Delta. This formula is computed for a range of sediment size from 0.15 to 1.5 mm.

$$K = 1.4e^{(-2.5D_{50})} \tag{3.35}$$

Kamphuis et al. (1978) linked the empirical parameter with the surf similarity parameter.

$$K = 0.7\zeta_b, \qquad where, \zeta_b = \frac{m}{\sqrt{H_{brms}/L_0}} \tag{3.36}$$

L_o is deep water wavelength, and H_{brms} is the root mean square breaker height.

Kamphuis et al. (1986) developed a sediment transport formula (3.37), and determined the calibration coefficient in term of sediment size and significant breaking wave height upon dimensional analysis.

$$\frac{Q_s}{\frac{1}{2}\rho\frac{H_{bs}^3}{T}\sin 2\alpha_{bs}} = \left(0.002\frac{H_{bs}}{D_{50}}\right)\frac{m}{(H_{bs}/L_o)^{\frac{1}{2}}} \quad where, K = \left(0.002\frac{H_{bs}}{D_{50}}\right), m = 1.8(H_b/D_{50})^{-0.5} \tag{3.37}$$

The correlation between the CERC and Kamphuis formula given in (3.38).

$$Q_s = 0.018\left(\gamma\frac{H_{bs}}{D_{50}}\right)^{\frac{1}{2}}Q_{s,CERC} \tag{3.38}$$

Where, D_{50} in meters.

King (2005) pointed out that the K in CERC formula based on

Kamphuis model can be calculated using the following equation.

$$K_{sig} = 0.022 \sqrt{\gamma \frac{H_{bs}}{D_{50}}}$$ (3.39)

Comparison among different formulas of the empirical parameter in terms of the grain size is presented in Figure 3.2. Also field data from different literature were added in Figure 3.2.

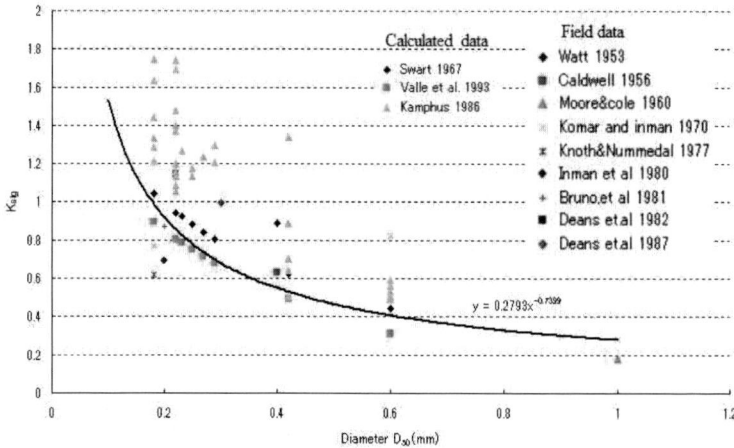

Figure 3.2: The empirical parameter (K) versus the grain size.

Power relation exists between K and the grain size. As the sediment size increases the calibration coefficient decreases, hence the sediment transport decrease. In considering the grain size effect in the longshore sediment transport rates, the calibration coefficients in Ozasa and Prampoton formula is calculated in term of the sediment size. The relation $K_1 \alpha\ (d)^{0.5}$ is introduced, as given by Kamphuis et al. (1986). $K_1 = A/(d)^{0.5}$, and $K_2 = 0.5\ K_1$. K_1 represent the mobility of the grains and A is calibrated using shoreline change data. If the grain size is larger, $K1$ and the longshore sediment transport rate Q become smaller, because sand with large grain size is difficult to move.

Model Structure

The model structure is shown in Figure 3.3. The input data consist of site specific wave climate data, morphological features and existing or planned structural information (such as groins, breakwaters, beach fill). The wave data, a set of wave events including deep water wave heights, periods, angles, corresponding closure depths and frequencies of occurrence for each wave direction per year, are obtained from either a wave history data or a wind climate study depending on the available type of data recorded by local meteorological stations. Another major input is the initial shoreline orientation and physical characteristics (such as median grain size in the surf zone, bottom slope or shape, berm height). The shoreline orientation is represented by an appropriate discretized shoreline. The last input data is the structural information such as the location and length of a groin or offshore distance of a detached breakwater. After all necessary input data is entered to the model, the shoreline is subject to waves and its evolution in time is observed.

The spatial and temporal changes in shoreline orientation are computed by two modules; wave transformation and morphology modules. The wave transformation module is where wave refraction, shoaling, diffraction and breaking calculations are conducted to evaluate local breaking conditions for all grid cells along the discretized shoreline. In the module, the deep water wave parameters are first transformed into breaking wave parameters and then the effects of diffracting sources are considered breaking wave heights and angles within the sheltered zone of structures are re-calculated. The output data are used in the morphology module to calculate local sediment transport rates and changes in the shoreline due to longshore sediment transport. The updated shoreline is used as input for the following time step and the same procedures continue until the end of simulation time.

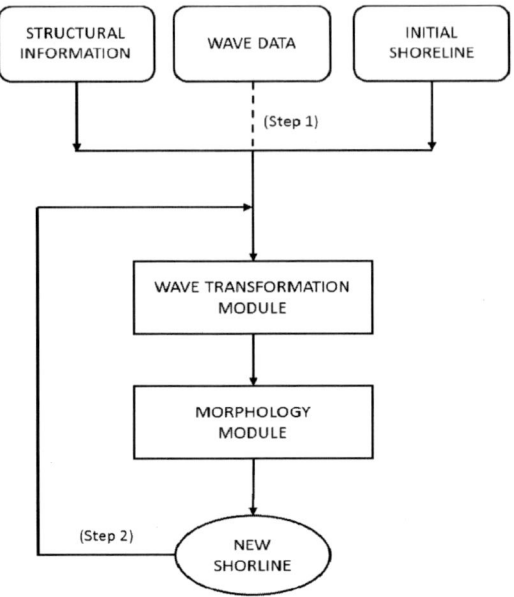

Figure 3.3: Model structure.

3.3.2. Grid System

The shoreline is discretized by a grid spacing of Δx and (N+1) calculation cells are formed, the position of shoreline is defined by (N+1) y coordinates of corresponding grids on the x axis (see Figure 3.5). For the calculation of y values at the next Δt time step, (N+2) Q values are needed. In a staggered grid system, the Q values are defined between two consecutive grid points whereas q values are defined at the grid points. The direction of sediment transport between two calculation cells is determined from the angle between wave crests and breaking bottom contour line such that if the waves are approaching the coast from left to right looking from shore-side, it is assumed to have a positive transport and a negative one for the waves approaching from right to left (see Figure 3.4).

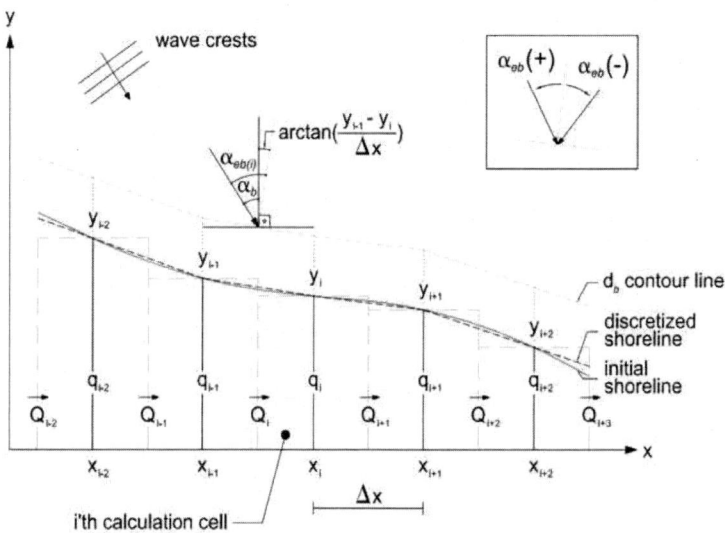

Figure 3.4: Grid system used for numerical modeling.

The numerical solution of equation of mass conservation (3.21) using an explicit scheme is as follows;

$$y_i^* = y_i + \frac{\Delta t}{(D_C + B)} \cdot \left[\frac{Q_i + Q_{i+1}}{\Delta x} + q_i \right] \qquad (3.40)$$

Where the prime (*) indicates a quantity at the next time step and Δx is the distance between two consecutive grids. The shoreline position of a calculation cell Δt time later is calculated by the difference in longshore transport rates entering in (Q_i) and out (Q_{i+1}) of the cell and net cross-shore gain or loss (q_i) over the active profile, ($D_{LT}+B$). If there are no sources or sinks, q_i is taken to be equal to zero. Further confining the sediment transport only to longshore sediment transport (cross-shore or other sediment transport mechanisms are not considered), then the seaward limit of the active profile is taken as the depth where no longshore sediment transport exists. To compute shoreline evolution and sand passing the tip of coastal structures (Şafak, 2006), this depth was defined as the limiting depth of longshore sediment transport (D_{LT}) and was related to wave breaking wave height (H_b). The key process in the longshore transport is given in equation (3.41) (Hallermeier, 1978) where H_b and T are the breaking wave height and associated period, respectively.

$$D_{LT} = 2.28\ H_b - 68.5 \left(\frac{H_b^2}{gT^2} \right) \qquad (3.41)$$

The i'th calculation cell in Figure 3.4 is drawn in detail in Figure 3.5 such that a section of shoreline with Δx length moves in cross-shore direction as backward or forward. Along the section, shoreline position is assumed to have a value of y, and y^* after Δt time.

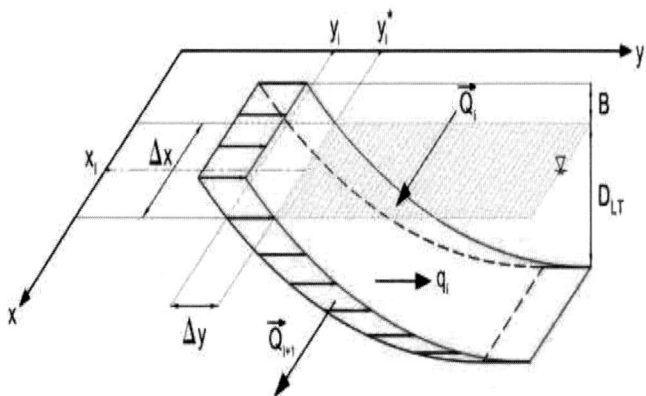

Figure 3.5: The i'th calculation cell in the numerical model.

To calculate sediment transport rates between two consecutive grids, the variables, introduced in the transport expression (Q), changing alongshore have also (N+2) values. In most of the cases, all the parameters included in the sediment transport rate equation except the breaking wave height, angle and bottom slope at breaking depth maybe assumed identical to all grids. Under changing shoreline conditions, breaking wave angle (α_b) measured between the bottom contours and wave crests also changes. Then, the local breaking wave angle called as the effective breaking wave angle

(α_{eb}) between two consecutive grid points is defined as;

$$\alpha_{eb(i)} = \alpha_b - \arctan\left(\frac{y_i - y_i}{\Delta x}\right) \qquad (3.42)$$

The breaking wave height alongshore the discretized shoreline is obtained from a wave transformation procedure including shoaling, refraction, breaking and diffraction mechanisms.

3.3.3. Boundary Conditions

In order to link the modeled shoreline to the outside environment and to solve equation of the longshore sediment transport rate, the boundary conditions for either y or Q at the two lateral ends of the beach are essential. Commonly applied lateral boundary conditions are Neumann or radiation boundary condition that represents the natural beach condition where the change in shoreline position is negligibly small and Dirichlet boundary condition which represents for an impermeable shore-normal barrier where sediment transport rate is equal to zero. Another type of boundary condition controls the cross-shore movement of shoreline as backward in case of seawall and as forward in case of tombolo formation behind detached breakwater.

The numerical solution of conservation of mass equation (3.40) requires (N+2) Q values for the calculation of (N+1) y values in time. Since the major parameter in the calculation of longshore sediment transport, the effective breaking wave angle is not defined at the two lateral ends, sediment transport rates at these ends are defined with respect to corresponding boundary condition. If the boundary holds for no significant shoreline change with time, $\frac{\partial y}{\partial t} = 0$, equation (3.21) then yields $\frac{\partial Q}{\partial x} = 0$ which may be expressed as $Q1 = Q2$ or $QN+2=QN+1$. If a complete shore-normal barrier that interrupts the longshore sediment movement, exists at one end of the shoreline, then this condition can be expressed as $Q1=0$ or $QN+2=0$. The complete barrier lateral boundary condition can also be introduced as an internal constraint at any interior location in the grid system, to represent the applications of groins or jetties as coastal defense measures depending on their capacity of wave energy absorption and sediment transport blocking. The amount of sediment blocked by the structure is related to the seaward extent of the groin with respect to the critical offshore distance that corresponds to the depth of closure (Dabees, 2000). The amount of wave energy absorption capacity, which also controls the sediment blocking, is expressed as the permeability of the groin. The details of the calculations of bypassing and permeability conditions of groins in the developed model are discussed extensively by Şafak (2006).

3.3.4. Stability Criterion

The determination of the sizes of time interval (Δt) and grid spacing (Δx) depends on the stability parameter (R_S) when other parameters are kept constant in equation (3.44). For small breaking wave angles, the stability parameter is given as:

$$Rs \leq 0.5 \tag{3.43}$$

Where

$$R_S = \left[\frac{Q \cdot \Delta t}{\alpha_{eb} \cdot (\Delta x)^2 \cdot (D_C + B)} \right] \tag{3.44}$$

The stability parameter gives an estimate of the numerical accuracy of the solution such that accuracy increases with decreasing values of Rs (Hanson and Kraus, 1986).

3.4. RMA Model

The RMA model is a finite element modeling system for two- and three-dimensional free surface flow developed by King et al. (1995). The model has two modules; a hydrodynamic module (RMA-10) and a water quality module (RMA-11). The hydrodynamic module (RMA-10) is based on the shallow water equations where the effects of Coriolis, surface wind stress are included. It simulates water levels and flows using an irregular grid system covering the area of interest. Input to the module comprises bathymetry, bed resistance coefficients, wind field, and hydrographic boundary conditions. The water quality module (RMA-11) can be used as an advection diffusion model by solving the equation of conservation of mass for dissolved or suspended substance to simulate the spread of such substances in an aquatic environment under the influence of the fluid transport. The advantage of the RMA model is that it uses an unstructured grid that allows the use of a coarse grid for the areas far away from the area of interest.

In order to describe stratified flow (due to salinity or thermal or suspended sediment influences) in three dimensions, the governing equations must describe velocity in all three Cartesian directions, water pressure and the distribution of salinity, temperature or suspended sediment throughout the system. These parameters form the principal dependent variables of the problem. A suitable set of equations can be achieved by combining the Reynolds form of the Navier Stokes equations, the volume continuity equation, the advection diffusion equation for salinity or temperature transport and an equation of state relating water density to salinity or temperature. For the purposes of this derivation the equations will be described in terms of salinity, an exactly parallel development is possible if temperature or suspended sediment is treated as the dependent variable. The governing equations may thus be written as:

Momentum:

Three equations for each Cartesian direction

$$\rho\{\frac{\partial u}{\partial t} + u\frac{\partial u}{\partial x} + v\frac{\partial u}{\partial y} + w\frac{\partial u}{\partial z}\} - \frac{\partial}{\partial x}(\varepsilon_{xx}\frac{\partial u}{\partial x}) - \frac{\partial}{\partial y}(\varepsilon_{xy}\frac{\partial u}{\partial y}) - \frac{\partial}{\partial z}(\varepsilon_{xz}\frac{\partial u}{\partial z})$$

$$+ \frac{\partial p}{\partial x} - \Gamma_x = 0 \qquad (3.45)$$

$$\rho\{\frac{\partial v}{\partial t} + u\frac{\partial v}{\partial x} + v\frac{\partial v}{\partial y} + w\frac{\partial v}{\partial z}\} - \frac{\partial}{\partial x}(\varepsilon_{yx}\frac{\partial v}{\partial x}) - \frac{\partial}{\partial y}(\varepsilon_{yy}\frac{\partial v}{\partial y}) - \frac{\partial}{\partial z}(\varepsilon_{yz}\frac{\partial v}{\partial z})$$

$$+ \frac{\partial p}{\partial y} - \Gamma_y = 0 \qquad (3.46)$$

$$\rho\{\frac{\partial w}{\partial t} + u\frac{\partial w}{\partial x} + v\frac{\partial w}{\partial y} + w\frac{\partial w}{\partial z}\} - \frac{\partial}{\partial x}(\varepsilon_{zx}\frac{\partial w}{\partial x}) - \frac{\partial}{\partial y}(\varepsilon_{zy}\frac{\partial w}{\partial y}) - \frac{\partial}{\partial z}(\varepsilon_{zz}\frac{\partial w}{\partial z})$$

$$+ \frac{\partial p}{\partial z} + \rho g - \Gamma_z = 0 \qquad (3.47)$$

Volume Continuity:

$$\frac{\partial u}{\partial x} + \frac{\partial v}{\partial y} + \frac{\partial w}{\partial z} = 0 \qquad (3.48)$$

Advection Diffusion:

$$\frac{\partial s}{\partial t} + u\frac{\partial s}{\partial x} + v\frac{\partial s}{\partial y} + w\frac{\partial s}{\partial z} - \frac{\partial}{\partial x}(D_x\frac{\partial s}{\partial x}) - \frac{\partial}{\partial y}(D_y\frac{\partial s}{\partial y}) - \frac{\partial}{\partial z}(D_z\frac{\partial s}{\partial z}) - \theta_s = 0$$

$$(3.49)$$

Equation of State:

$$\rho = F(s) \qquad (3.50)$$

Where, (u,v,w) are velocities in the Cartesian directions, ρ is the density, g is the acceleration of gravity, p is the water pressure, ε_{xx} is the turbulent eddy coefficients, Γ is the external tractions that operate on the boundaries or on the interior, S is the salinity, D_x is the eddy diffusion coefficients for salinity and θ_s is the source/sink for salinity.

In general, the geometric system over which equations 3.45 through 3.50 must be solved, varies with time, that is, the water depth h varies during the simulation. In order to develop an Eulerian form for the solution it is desirable to transform this system to one that can be described with a constant geometric structure. In the earliest development of this model (King 1982), the generalized system with irregular water surface elevation and bottom profile was transformed to a regular system with unit vertical dimension.

This transformation is conventionally referred to as the sigma transformation. The governing equations were then modified to reflect this new coordinate system and the system solved with constant geometric dimensions. In a later analysis of this method, King (1985 and 1995) pointed out, that at locations where a sharp break in bottom profile occurs the transformation is not unique and momentum in the component directions may not be correctly preserved. As an alternative a modified transformation was presented that preserved the bottom profile as defined, but transformed the water surface to a constant elevation. RMA-10 uses this modified transformation. When this transformation is applied, the vertical velocity component at the bed must be adjusted so that overall continuity is maintained. That is, a vertical velocity component is computed so that for any node on the stream bed any flow leakage out of the system over a defined tributary area is balanced by flow leakage into the system over the same tributary area.

Chapter 4
Numerical Simulation

4.1. General

Numerical simulation has been implemented by using the Surface Water Modeling System (SMS-10.1). The numerical model was applied to investigate various configurations of the perched beach including submergence ratio of the breakwater, groin with/without gap, the gap width/location and emerged/submerged groins. These configurations have been compared from the point of view of wave height, currents velocities, flushing rates and shoreline changes to develop general guidelines for the design of similar constructions.

4.2. Model Input

A study has been conducted on the use of perched beach for protecting a strip of the beach from severe wave attack while keeping reasonable flushing rates condition and minimum shoreline changes due to the structure. The layout of the proposed perched beach is shown in Figure 4.1. Simulations using SMS model have been made for 1500m along the shore line and extends for about 700m normal to the shoreline to limit the boundary effects. The perched beach generally has two bounding groins and shore parallel breakwater at water depth equal to 3.5m. The groins extend for 120m normal to the shoreline and have some opening (clear gap) in its body and the distance between groins is 200m. The average effective grain size of the soil on-site is found to be 0.25mm, the average berm height is computed and found to be 2m and the closure depth is considered as 8m.

The bathymetric survey has been to run the model presented in Figure 4-2. It can be seen that the bed levels go as deep as about 20m within 700m from the existing shoreline with an overall average slope of 1:35. It can be observed that the bed has a nearly uniform slope at bed levels less than 3m below MSL.

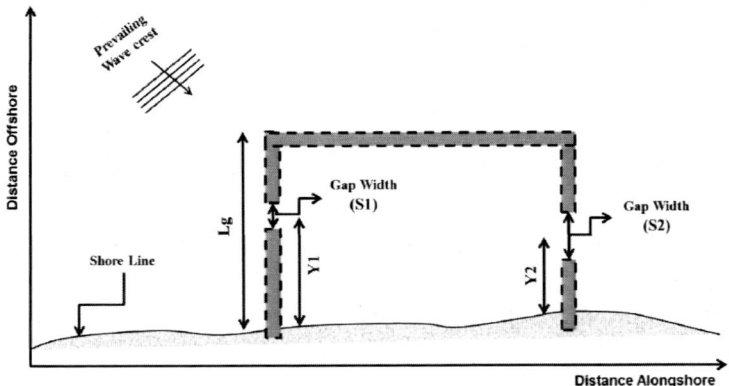

Figure 4.1: Layout of the proposed structure.

Where, Lg is the groin length, S1 is the gap width in western groin, S2 is the gap width in eastern groin, Y1 is the distance from the centerline of western gap to the shoreline and Y2 is the distance from the centerline of eastern gap to the shoreline.

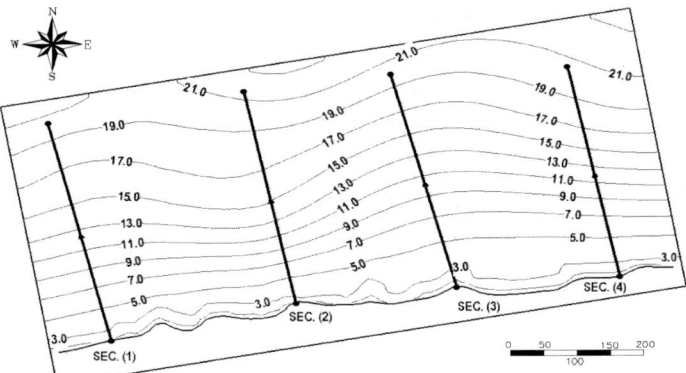

Figure 4.2: Bathymetric survey and location of transects at station.

The wave data has been analyzed and input to the numerical simulation model (SMS). SMS considers the existing shoreline as the baseline and the normal to it as a false North (direction=0 for the normal direction to the shoreline and angle is positive anti-clockwise). The distribution of the

- 62 -

deepwater wave energy (m2/Hz) versus the direction angle measured from the normal to the shore line during the occurrence of maximum incident wave height is shown in Figure 4-3. It can be observed that most of the wave energy is centered about the normal to the shoreline and the maximum is reached at about 10 degrees to the normal on the shoreline. Moreover, the spectral analysis of deepwater wave data is given in Table 1 as computed by SMS.

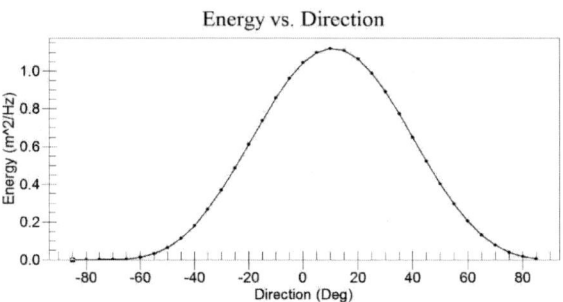

Figure 4.3: Wave energy distribution versus direction during the maximum incident wave height in deep water.

Table 4.1: Spectral analysis of wave data computed by SMS.

Index	Frequency of Occurrence	Hs (m)	Tp (sec)	Direction Angle (degrees)	Index	Frequency of Occurrence	Hs (m)	Tp (sec)	Direction Angle (degrees)
10	10	0.125	3.842	-62.206	160	43	1.189	7.138	-15.297
20	32	0.125	3.763	-39.289	170	134	1.194	7.117	14.52
30	66	0.128	3.771	-9.99	180	9	1.613	7.73	-9.143
40	152	0.126	3.764	19.968	190	55	1.66	7.819	13.387
50	12	0.185	4.258	-62.206	200	2	1.865	8.785	-36.05
60	180	0.321	4.932	-41.594	210	4	1.84	8.132	-20.245
70	275	0.324	4.826	-9.399	220	23	1.914	8.122	13.002
80	510	0.304	4.744	18.837	230	1	2.063	9.02	-36.05
90	29	0.548	5.812	-37.025	240	9	2.339	8.571	-10.143

100	61	0.593	5.791	-8.588	250	67	2.501	8.623	12.435
110	135	0.588	5.792	16.986	260	8	3.247	9.436	-13.981
120	68	0.696	6.356	-39.121	270	26	3.317	9.423	11.593
130	84	0.822	6.392	-11.593	280	2	4.077	9.93	11.593
140	185	0.815	6.368	15.874	290	6	4.506	10.21	10.758
150	3	1.066	7.487	-41.503					

The wave data and bathymetric survey described in the aforementioned sections have been used to run the model for the current configurations (see Table 4.2) and the results have been presented, analyzed and discussed. The results and analysis provide general guidelines for the use of perched beach in coastal resorts that can be applied to wide range of wave climate.

Table 4.2: Input parameters for the numerical model (SMS).

Run No.	Breakwater		Western Groin			Eastern Groin			Notes	
	Crest Level	d/h	Crest Level	(S1/Lg)	Y1	Crest Level	(S2/Lg)	Y2		
01	-0.50	0.84	-0.50	---	---	-0.50	---	---	No gaps (Study the submergence ratio)	
02	-0.90	0.72	-0.90	---	---	-0.90	---	---		
03	-1.25	0.62	-1.25	---	---	-1.25	---	---		
04	-0.50	0.84	+2.0	---	---	+2.0	---	---	Submerged breakwater without gaps	Emerged groins
05	-0.50	0.84	+2.0	0.10	0.75 Lg	+2.0	0.20	0.25 Lg	Submerged breakwater with gaps	
06	-0.50	0.84	-0.50	0.05	0.75 Lg	-0.50	0.20	0.25 Lg	Effect of western groin width	Study the effect of the eastern groin gap location
07	-0.50	0.84	-0.50	0.10	0.75 Lg	-0.50	0.20	0.25 Lg		
08	-0.50	0.84	-0.50	0.15	0.75 Lg	-0.50	0.20	0.25 Lg		
09	-0.50	0.84	-0.50	0.20	0.75 Lg	-0.50	0.20	0.25 Lg		
10	-0.50	0.84	-0.50	0.05	0.75 Lg	-0.50	0.20	0.50 Lg	Effect of western groin width	
11	-0.50	0.84	-0.50	0.10	0.75 Lg	-0.50	0.20	0.50 Lg		
12	-0.50	0.84	-0.50	0.15	0.75	-0.50	0.20	0.50		

No					Lg			Lg		
13	-0.50	0.84	-0.50	0.20	0.75 Lg	-0.50	0.20	0.50 Lg		
14	-0.50	0.84	-0.50	0.05	0.75 Lg	-0.50	0.20	0.75 Lg		
15	-0.50	0.84	-0.50	0.10	0.75 Lg	-0.50	0.20	0.75 Lg	Effect of western groin width	
16	-0.50	0.84	-0.50	0.15	0.75 Lg	-0.50	0.20	0.75 Lg		
17	-0.50	0.84	-0.50	0.20	0.75 Lg	-0.50	0.20	0.75 Lg		
18	-0.50	0.84	-0.50	0.10	0.75 Lg	-0.50	0.10	0.25Lg	Effect of western groin location	
19	-0.50	0.84	-0.50	0.10	0.75 Lg	-0.50	0.10	0.50 Lg		
20	-0.50	0.84	-0.50	0.10	0.75 Lg	-0.50	0.10	0.75 Lg		Study the effect of the eastern groin gap location
21	-0.50	0.84	-0.50	0.10	0.50 Lg	-0.50	0.10	0.25 Lg	Effect of western groin location	
22	-0.50	0.84	-0.50	0.10	0.50 Lg	-0.50	0.10	0.50 Lg		
23	-0.50	0.84	-0.50	0.10	0.50 Lg	-0.50	0.10	0.75Lg		
24	-0.50	0.84	-0.50	0.10	0.25 Lg	-0.50	0.10	0.25 Lg	Effect of western groin location	
25	-0.50	0.84	-0.50	0.10	0.25 Lg	-0.50	0.10	0.50 Lg		
26	-0.50	0.84	-0.50	0.10	0.25 Lg	-0.50	0.10	0.75 Lg		

Where, Lg is the groin length, S1 is the gap width in western groin, S2 is the gap width in eastern groin, Y1 is the distance from the centerline of western gap to the shoreline and Y2 is the distance from the centerline of eastern gap to the shoreline.

4.3. Results and Analysis

The numerical models were applied to investigate various configurations of the perched beach, as follows:

- **Breakwater height(Submergence ratio).**
- **Groin Configurations (groin height - gab location/width).**

These configurations have been compared from the point of view of wave height, currents velocities, flushing and shoreline changes to develop general guidelines for the design of similar constructions. Different wave conditions were tested, as well, for the different configurations.

4.3.1. Breakwater Height

The value of the transmitted wave height is compared for different ratios of breakwater height to the water depth (0.84, 0.72 and 0.62), as shown in Figure 4.4. It is observed that wave breaking takes place over the submerged breakwater and small waves are transmitted to the protected area. These figures confirm that a partial standing wave is formed in front of the breakwater and nonlinear wave damping takes place over it and a smaller wave is transmitted to the onshore side. The results also suggest that wave energy dissipation depends greatly on breakwater height. It is shown that the higher the breakwater is, the lower the transmitted but the slightly higher the reflected wave energies are.

It has been found that the transmitted wave height is less than 0.60m for d/h=0.84, which mean that the site can be used for recreational purposes for most of the summer season. On the other hand, the transmitted wave height is more than 0.6m for d/h=0.72 and 0.62.So, these wave conditions are considered as non-comfortable swimming conditions, as stated by Saski et al (1975) and hence these Submergence ratios (0.72 and 0.62) do not provide safe swimming conditions.

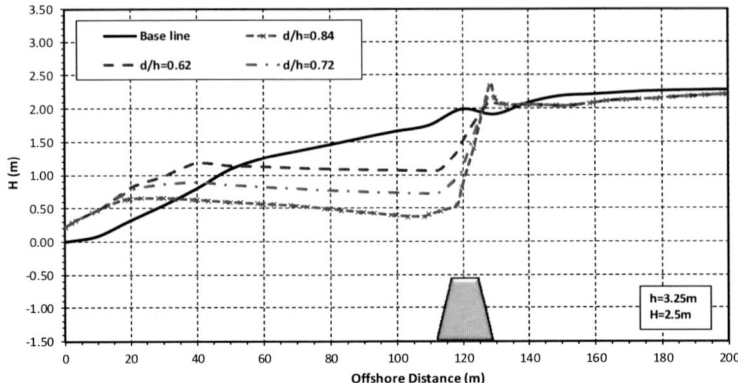

Figure 4.4: Effect of d/h on the wave height at the centerline of the perched beach (H=2.50m).

Computations have also been made for the transmission coefficient (Kt) using various well-known formulae available in literature. This includes the following formulae:

Computation of Kt using van der Meer and Daemen formula (1994):

Van der Meer and Daemen (1994) computed the value of Kt using the following equations:

$$K_t = a \frac{R_c}{D_{ns0}} + b \qquad (4.1)$$

Where

$$a = \left(0.031 \frac{H_i}{D_{n50}}\right) - 0.24$$

For conventional breakwaters, the value of b is computed as follows:

$$b = 0.51 - 5.42 S_{op} + 0.0323 \frac{H_i}{D_{n50}} - 0.0017 \left(\frac{B}{D_{n50}}\right)^{1.84}$$

For reef breakwaters, the value of b is computed as follows:

$$b = 0.85 - 2.6 S_{op} - 0.05 \frac{H_i}{D_{n50}}$$

Where S_{op} is the local wave steepness (2 H/gT^2) and R_c is the crest freeboard relative to the still water depth (h-d).

For conventional breakwaters, the value of K_t has a minimum of 0.075 and a maximum of 0.75. On the other hand, for reef breakwaters, the value of K_t ranges from 0.15 to 0.6.

Computation of Kt using d'Angremond et al formula (1996):

d'Angremond et al formula (1996) developed empirical formulae for estimating the coefficient of wave transmission as given by equation (4.2). The value of Kt has a minimum of 0.075 and a maximum of 0.8. For B/Hi < 10, the coefficient is computed as follows:

$$K_t = -0.4 \left(\frac{R_c}{H_i}\right) + 0.64 \left(\frac{B}{H_i}\right)^{-0.31} \left(1 - e^{(-0.5\,\zeta)}\right) \qquad (4.2)$$

Where ζ is the breaker parameter $= \tan \ / (S_{op})^{0.5}$

Computation of Kt using van der Meer et al. (2005):

For B/Hi > 10, van der Meer et al. (2005) revised the equation of d'Angremond et al (1996) to better match the experimental measurement as follows:

$$K_t = -0.35 \left(\frac{R_c}{H_i}\right) + .51 \left(\frac{B}{H_i}\right)^{-0.65} \left(1 - e^{(-0.41\zeta)}\right) \qquad (4.3)$$

$$K_{tu} = -0.006 \left(\frac{B}{H_i}\right) + 0.93 \qquad (4.4)$$

Where $\tan$ is the seaward slope of the structure K_{tu} is the upper limit of the coefficient K_t, and K_{tl} is the lower limit of the coefficient K_t which is 0.05.

For smooth LCS having $\zeta_{op} < 3$, the value of k_t is computed by van der Meer et al. (2005) as follows:

$$K_t = -0.30 \left(\frac{Rc}{H_i}\right) + \left(0.75 \left(1 - e^{-0.5\zeta_{op}}\right)\right) (COS(\beta))^{2/3} \qquad (4.5)$$

The value of k_t has a minimum of 0.075 and a maximum of 0.8 and is the angle that the orthogonal makes with the normal to the breakwater, i.e., =0 for normal wave incidence. Equation (4.5) is valid for values of $1 < \zeta_{op} < 3$, $\beta = (0 - 70)$ degrees and $B/H_i = 1 - 4$.

Computation of Kt using Tanaka (1976):

Goda and Ahrens (2008) simulated the design diagram developed by Tanaka (1976) for a wide range of laboratory data on model Low Crested

Structures (LCS) of reef-type by fitting the design curves to equation (4.6). The latter equation is referred to hereinafter as Tanaka (1976) since it reproduces the transmission coefficient as presented in Tanaka's design diagram.

$$K_t = \max\left(0, 1 - \exp\left[a\left(F/H_i - F_0\right)\right]\right) \tag{4.6}$$

Where

$$a = 0.248\, exp\left(-0.384\, \ln\left(B_{eff}/L_0\right)\right)$$

$F_0 = 1.0$ for $D_{eff} = 0$ (impermeable structure)

$F_0 = \max(0.5, \min(1.0, H/D_{eff}))$ for $D_{eff} > 0$ (permeable structure)

Where D_{eff} is the effective diameter of the rubble mound and is also denoted as D_{50}, B_{eff} is the top width of the crown to insert in Tanaka equation.

Computation of Kt using Goda and Ahrens (2008):

Furthermore, Goda and Ahrens (2008) considered that the coefficient of transmission (K_t) to be the summation of the coefficients of transmission through (K_{tthru}) and over (K_{tover}) LCS. They used the formula of Numata (1975) developed for estimating the transmission coefficient of waves passing through a sloped mound to compute the value of K_{tthru} as follows:

$$K_{tthru} = 1 / [\ 1 + C\ (\ H/L)^{0.5}\ ]^2 \ : \ C = 1.135(\ B_{eff}/D_{eff})^{0.65} \tag{4.7}$$

They also used the formula of Tanaka (1976), to compute the value K_{tover} but computed the value of B_{eff} for submerged breakwaters as follows:

$$B_{eff} = (9 \text{ x crest width + bottom width}) / 10 \tag{4.8}$$

Thus, the value of K_t is:

$$K_t = K_{tthru} + K_{tover} \tag{4.9}$$

The values of the coefficient of transmission either computed by the empirical formulae or the results of the SMS model have been presented in Figure 4.5. It has been found that the value of Kt for d/h=0.84 is 24% as per the SMS model, but it is 22%, 38%, 45%, 52% and 68% as computed by van der Meer et al (2005), d'Angremond et al (1996),van der Meer and Daemen

(1994),Tanaka (1976) and Goda and Ahrens (2008), respectively. It can be concluded that the SMS model results agree reasonably with those computed by van der Meer et al (2005) and d'Angremond et al (1996), but Tanaka (1976), van der Meer and Daemen (1994) and Goda and Ahrens (2008) show much higher values. Thus, only the results of van der Meer et al (2005) and d'Angremond et al (1996) are considered applicable to the current study. Generally, it can be stated that the maximum transmitted wave height in summer varies from 0.55m to 0.95m as computed by the SMS model, van der Meer et al (2005) and d'Angremond et al (1996) formulae.

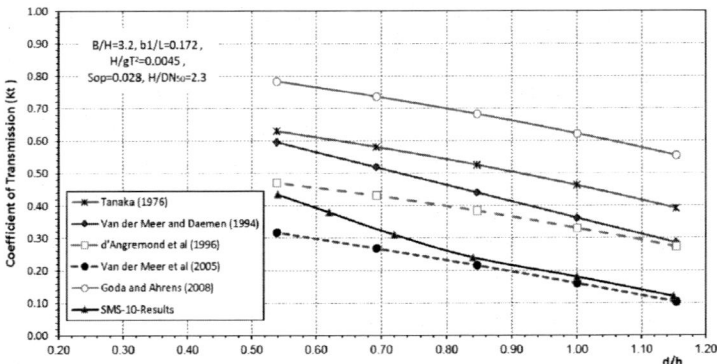

Figure 4.5: Computed values of the transmitted wave height using various empirical formulae and SMS for various values of d/h.

The value of the cross and longshore currents inside the perched beach is compared for different ratios of breakwater height to the water depth (0.84, 0.72 and 0.62), as shown in Figures4.6-4.7. It has been found that the currents inside the perched beach is less than 0.50m/s for d/h=0.84. On the other hand, the currents inside the perched beach is more than 0.5m/s for d/h=0.72 and 0.62.So, it can confirmed that the higher the breakwater is, the lower offshore currents are. Moreover, it can be observed that the effect of the submergence ratio (d/h) on the rip current is small compare to its effect on the long shore currents.

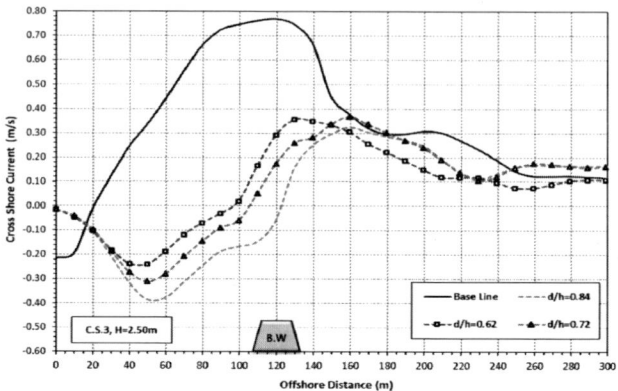

Figure 4.6: Rip current velocity along the perched beach for various values of d/h.

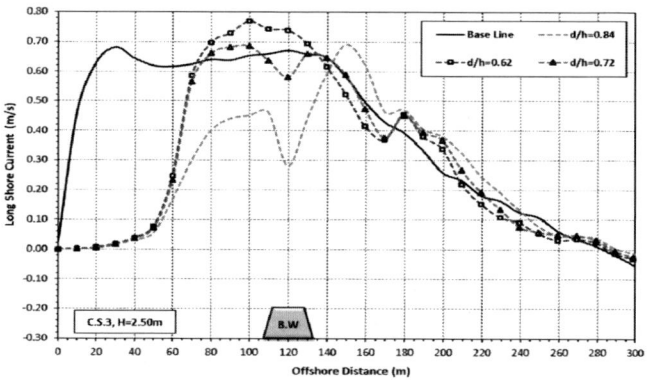

Figure 4.7: Long shore current velocity along the perched beach for various values of d/h.

Computations have been made using GENSIS program within the package of SMS model for shoreline locations. The longshore sediment transport has been calculated based on Ozasa and Brampton formula in SMS model. The net transport has been computed over three years of simulation for different ratios of breakwater height to the water depth (0.84, 0.72 and 0.62), as shown in Figure 4.8. It has been found that the submergence ratio has significant effect on the shoreline changes and the accretion is generally slightly larger than erosion, as shown in Figure 4.9.

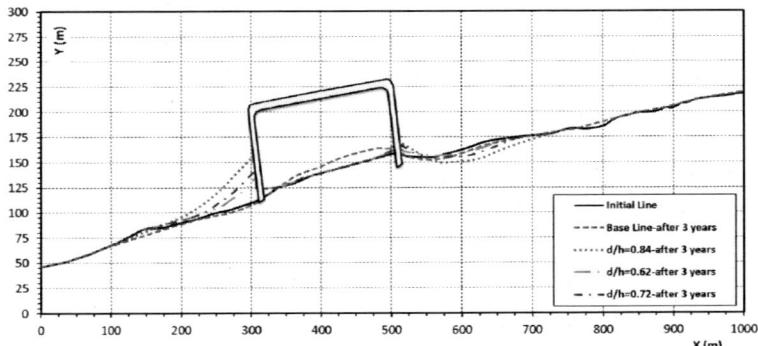

Figure 4.8: Shoreline changes around the perched beach for various submergence ratios after 3 years of construction.

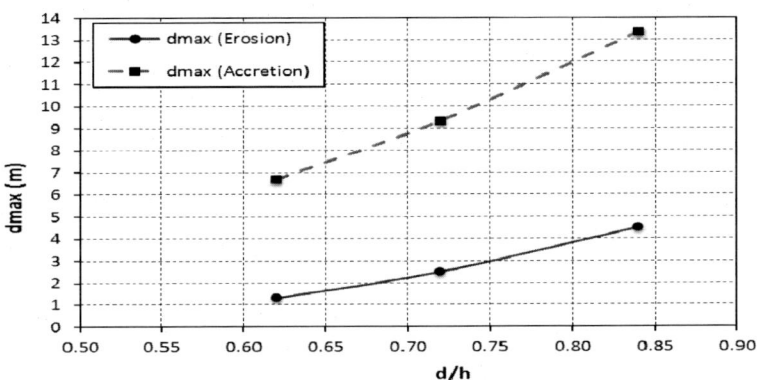

Figure 4.9: Computed maximum length of erosion/accretion along the shoreline for various submergence ratios after 3 years of construction.

Further computations have been made for the flushing rate using RMA models based on the relation between the tidal range and the structure height. The flushing time is measured as the time for the concentration to drop down to 10% of its initial value (Goshow et al., 2008).According to Goshow et al.

(2008) and shown in Figure4.10, the suitable flushing time for the proposed perched beach area is 5 days.

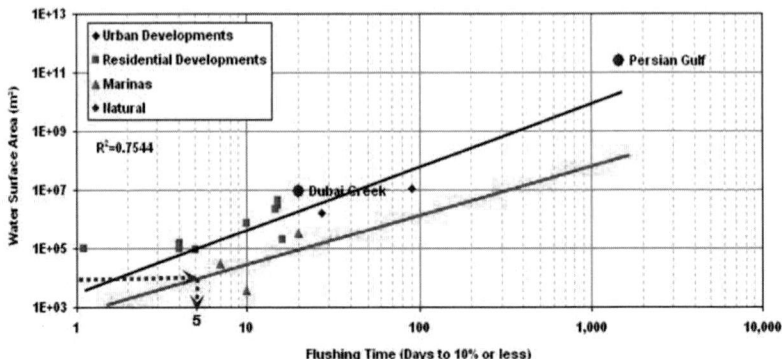

Figure 4.10: Relation between water surface area and the flushing time for water bodies (Goshow et al., 2008).

In order to assess the time scale associated with dilution of pollutants, an initial tracer concentration of 100% was introduced in the numerical model inside the perched beach. The concentration of inlet water at the boundaries is assumed to be zero. Figure 4.11 provides the five stations used to monitor the dilution/dispersion of a conservative material. Figures 4.12-a through 4.12-b show the concentration patterns at different times for d/h=0.84 and d/h=0.72.

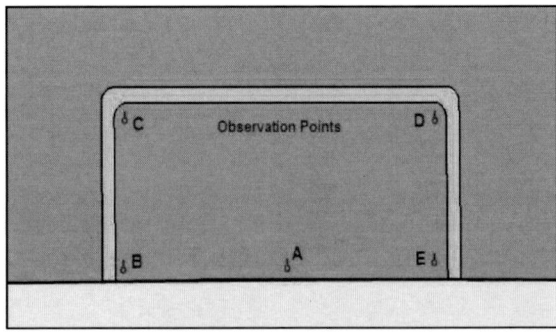

Figure 4.11: Locations used to plot time series of concentration for conservative material.

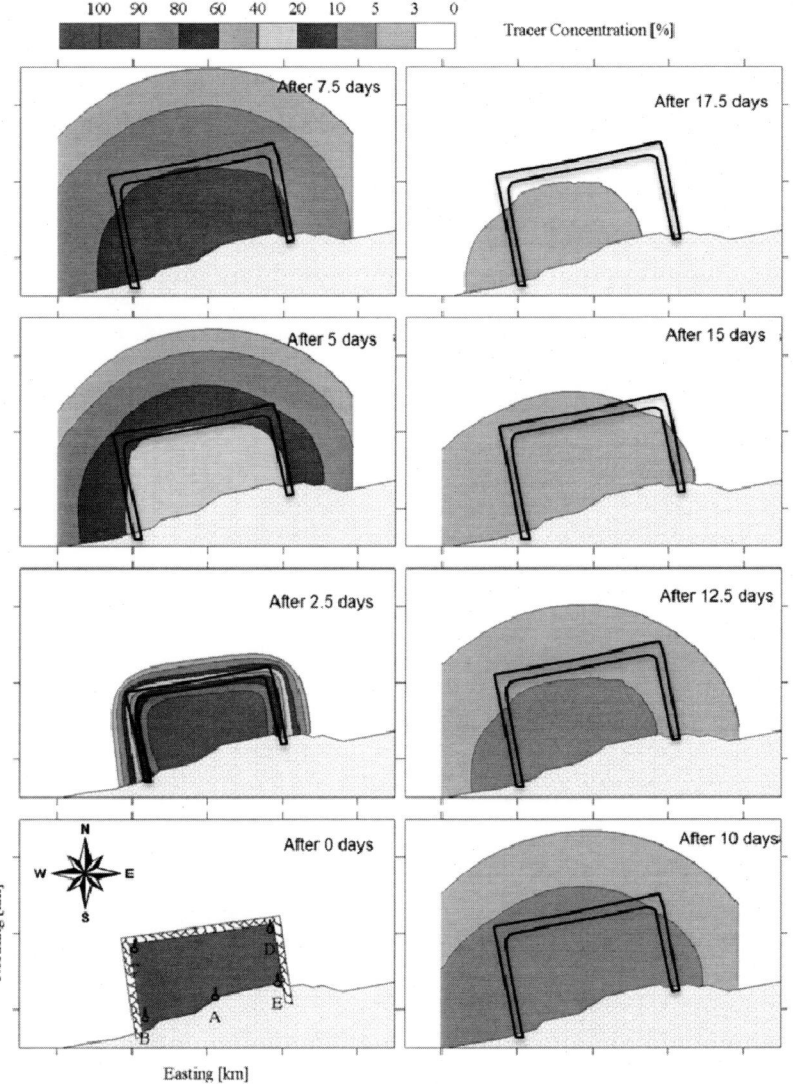

Figure 4.12-a: Concentration of tracer material at different times (d/h=0.84).

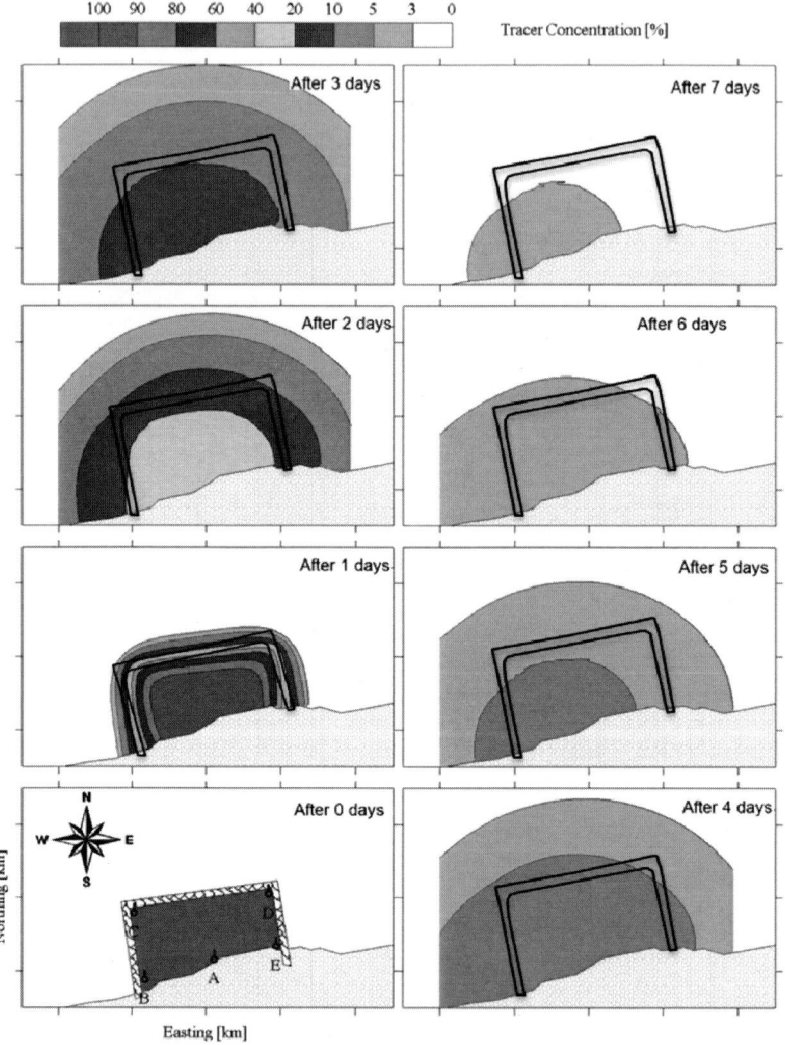

Figure 4.12-b: Concentration of tracer material at different times (d/h=0.72).

Variations in concentrations with time are plotted in Figure 4.13 for all locations. It can be observed that water stagnation occurs at station (B) and the flushing process is slow, while the flushing process at station (D) is the fastest inside the perched peach. So, special attention should give to station (B) when studying the flushing rates inside a perched peach.

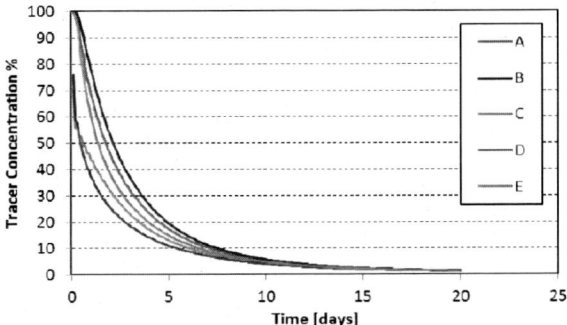

Figure 4.13: Time series for concentration of tracer material at different locations (d/h=0.84).

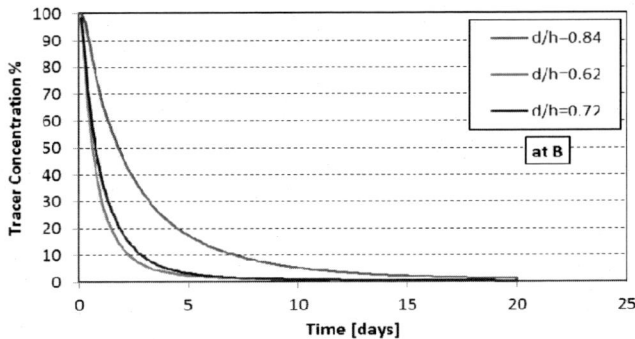

Figure 4.14:Time series for concentration of tracer material for various submergence ratios at station (B).

Figure 4-14 shows the time series for concentration of tracer material for various submergence ratios at station (B). It is shown that the higher the breakwater is the higher the flushing time. It can be seen that the flushing process is high in case of d/h= 0.72 and 0.62. The time associated with reduction of the average concentration to 10% of its initial value is about 2-3 days. This value is less than the target value of 5 days. It can be seen that the flushing process is slow in case of d/h= 0.84. The flushing time for d/h=0.84 has been found to be 7.5 days. This value exceeds the target value of 5 days. It is known that the long flushing time will lead to deterioration in water quality unless appropriate interference has been applied. This includes lowering the crest of the northern breakwater to be at less than 0.72 of the water depth or introducing a suitable gap in the groins, especially the east one. However, attention shall be made that reducing the height of the northern breakwater would eventually lead to higher waves and fast currents to be generated inside the perched beach as explained earlier.

4.3.2. Groin Configurations

In this section investigation have been made for the effect of the groin configuration as represented by groin height and gap location/width along the groin. The values of the transmitted wave height have been compared for various cases of groins height (emerged/submerged) and with/without gap, as shown in Figure 4.15. It has been found that there is no significant effect on the wave height at the centerline of the perched beach due to the gaps in groin and the emerged/submerged groins for the cases investigated.

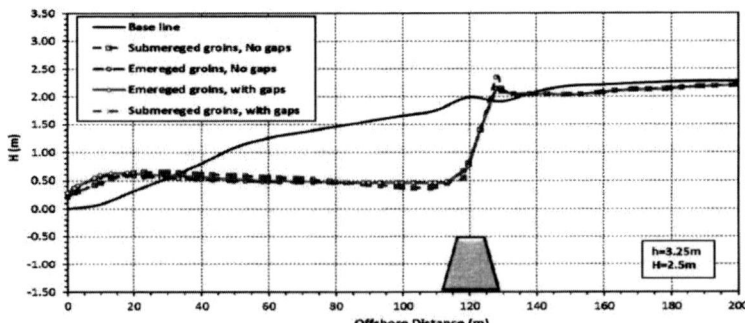

Figure 4.15: Effect of the gaps in groin and the emerged/submerged groins on the wave height at the centerline of the perched beach (H=2.50m).

The emerged/submerged groins have significant effect on the cross and longshore currents inside the perched beach. On the other hand, the gaps in groin has no effect on the cross and longshore currents inside the perched beach for the studied cases, as shown from Figures 4.16-4.17.

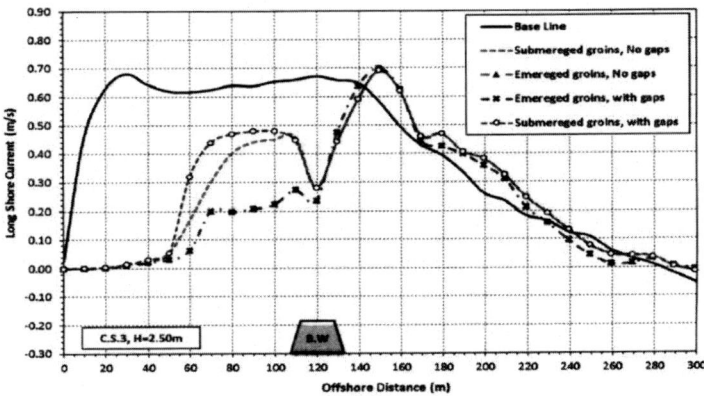

Figure 4.16: Effect of the gaps in groin and the emerged/submerged groins on the long shore current velocity along the perched beach(H=2.50m).

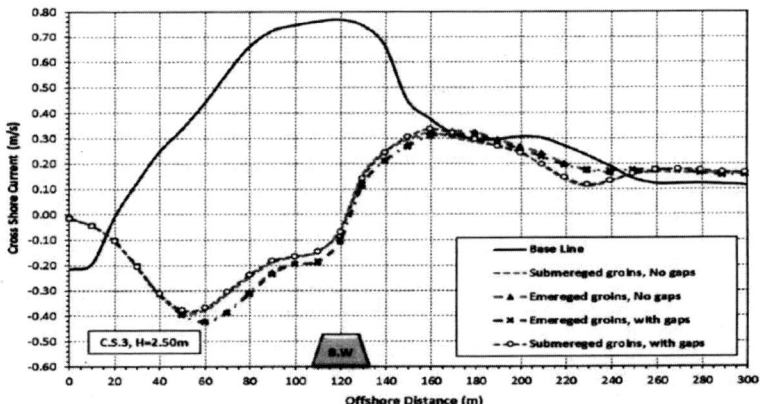

Figure 4.17: Effect of the gaps in groin and the emerged/submerged groins on the rip current velocity along the perched beach(H=2.50m).

The effects of gaps in the groins and the height of the groins on the shoreline changes around the perched beach after 3 years of construction are presented in Figure 4.18. It can be observed that the height of the groins has significant effect on the shoreline changes along the up drift and down drift of the perched beach. Accretion is slightly slower in case of the gap than in case of no gap, this is due to the fact that sand moves through the gap from the up drift side to the down drift one. Thus, further investigations have been made for the gap width/location.

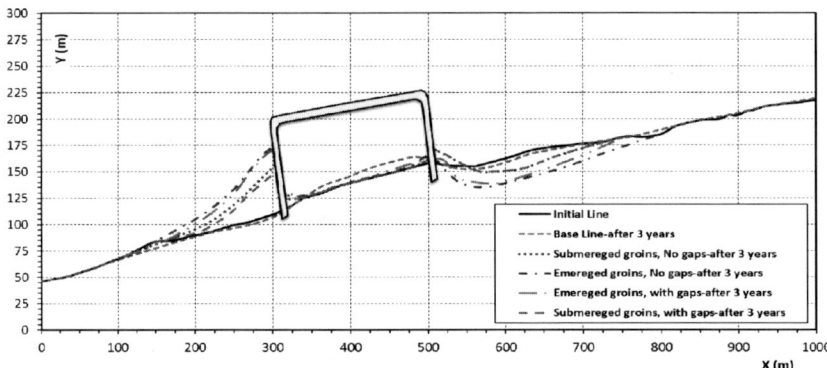

Figure 4.18: Effect of the gaps in groin and the emerged/submerged groins on the shoreline changes around the perched beach after 3 years of construction.

Figure 4.19 shows the shoreline changes along the coast due to various sizes of the western gap after three years from the complete construction of the perched beach. It can be noticed that considerable changes have occurred along the shoreline in the vicinity of the perched beach. These changes diminish within three times the length of the groin on both sides of the perched beach. It can be observed that accretion occurs at the up-drift side and erosion develops on the down drift one. It can be noticed that accretion on the up-drift side decreases with increasing the western gap width.

Also, erosion decreases on the down-drift side with increasing the western gap width. Moreover, accretion occurs inside the perched beach and increases as the western gap width increases. This is due to the fact that sand moves through the gap from the up drift side. Part of the sediment deposits inside the perched beach and the rest moves through the eastern gap to the down drift zone, and hence erosion is reduced at the down drift side. It can be stated that if the gap width is 0.1Lg very limited sedimentation is developed inside the perched beach after three years from construction, while larger gap width causes accelerated accretion inside the perched beach.

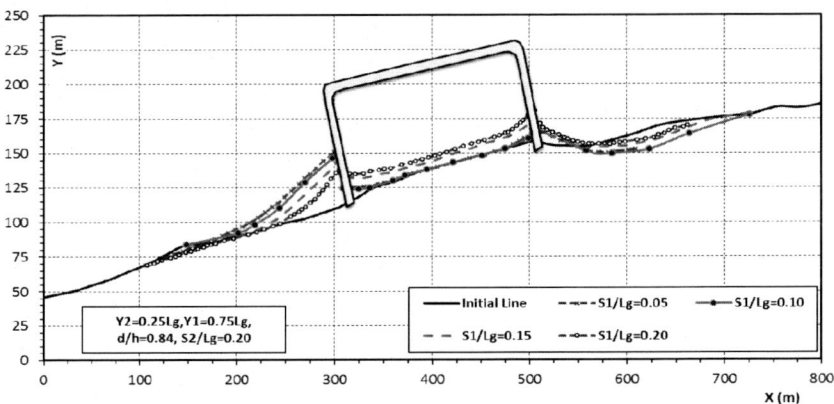

Figure 4.19: Computed shoreline changes along the coast for various gap widths after 3 years of construction.

Figure 4.20 shows the effect of the eastern gap location on the shoreline changes along the coast after 3 years of construction. It can be observed that accretion inside the perched beach and erosion on the down drift decreases when the gap is closer to the shore. This is due to the fact that sand moves shortly through the gap from the up drift side allowing part of the sediment to be deposited inside the perch beach while the remaining part moves to the down drift zone and nourish it. This mechanism becomes more evident as the gap in the western groin is closer to the shore.

Based on the mechanism of sand movement at the up drift side, the rate of accretion is commonly is the heights shortly after construction of the perched beach causing blocking of any gap constructed close to the shore line and hence preventing further sand movement through the gap. This also causes the loss of gap function in allowing current flowing into the perched beach which is the prime cause of flushing the water body. Based on the aforementioned reasons only the case of gap constructed along the western groin away from the shoreline has been considered in the present study. A gap constructed at 0.75 of the groin length from the shoreline has been considered in the present study and is considered in the analysis of various alternatives. Additionally the gap width has been fixed at 0.1 Lg for the analysis of various alternatives.

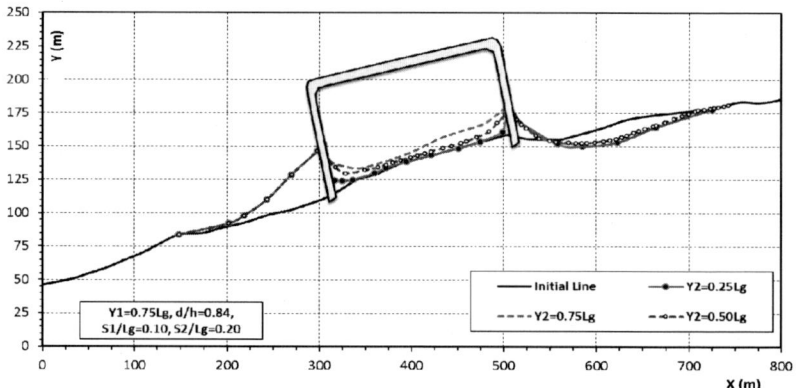

Figure 4.20: Effect of the eastern gap location on the shoreline changes along the coast after 3 years of construction.

Figures 4.21-4.22 show the concentration patterns in case of emerged groins with/without gaps. Figure 4-23 shows the time series for concentration of tracer material for various cases of groins height (emerged/submerged) and with/without gap at station (B). The flushing times are about 5 days, 7.5 days, 10 days and 12.5 days in cases of submerged groins with gaps, submerged groins without gaps, emerged groins with gaps and emerged groins without gaps, respectively. It can be observed that the height of the groins and the gaps have significant effect on the flushing process. Flushing process is slightly higher in case of the submerged groins than in case of emerged groins. Also, flushing process is

- 81 -

slightly higher in case of the gap than in case of no gap, this is due to the fact that the gaps allow water circulation and flushing in the near shore zone. Thus, further investigations have been made for the gap width/location.

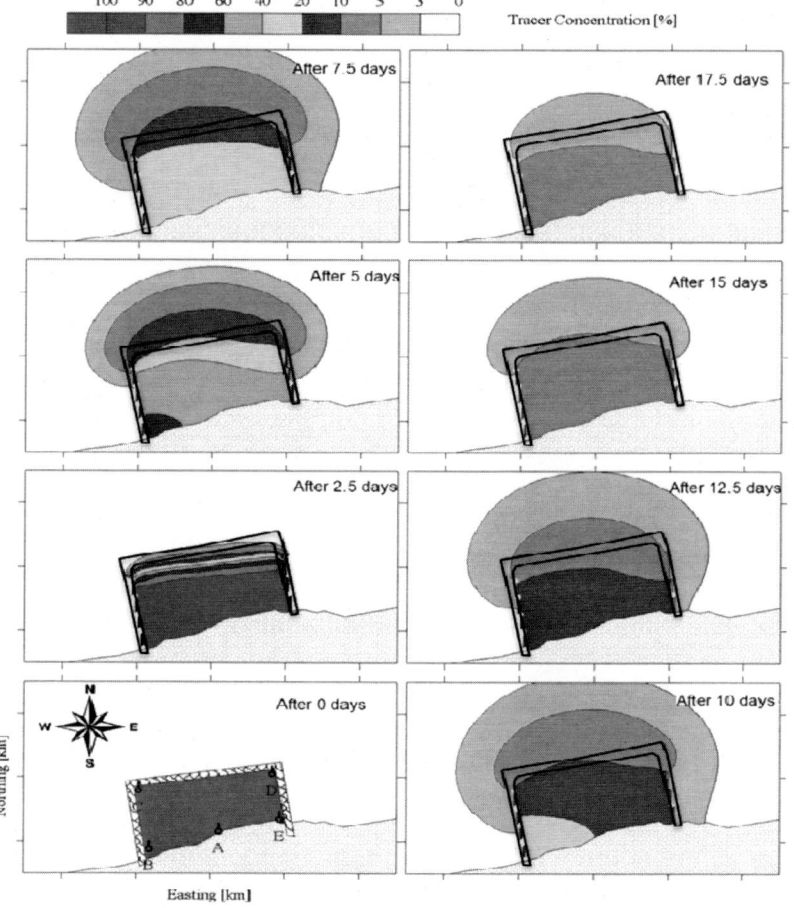

Figure 4.21: Concentration of tracer material in case of emerged groins without gaps.

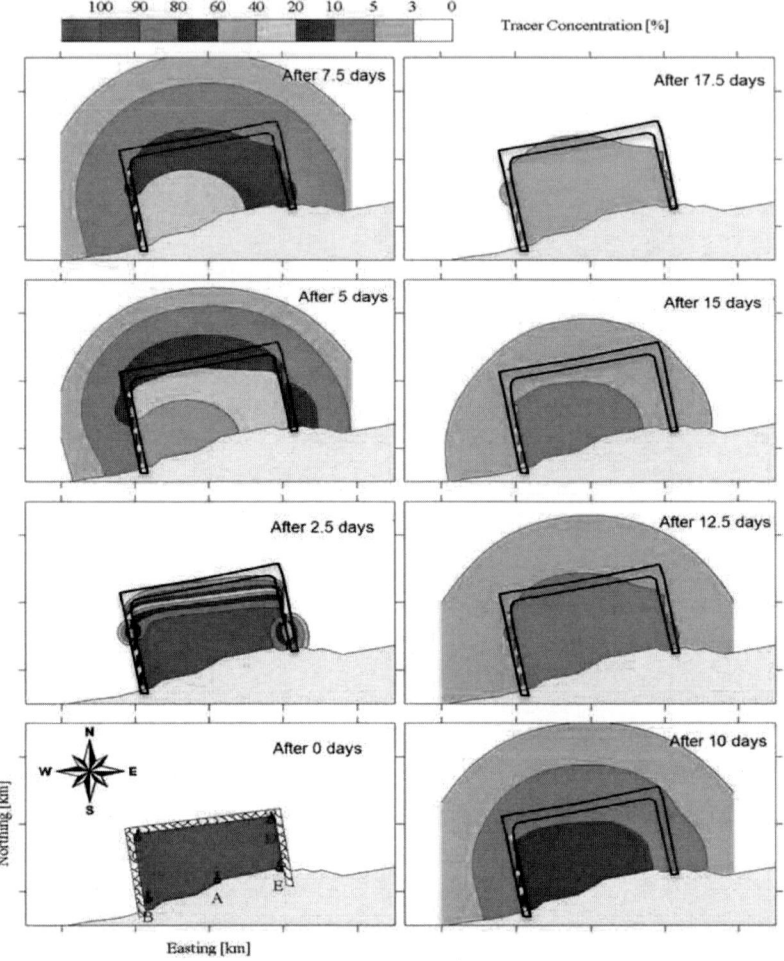

Figure 4.22: Concentration of tracer material in case of emerged groins with gaps.

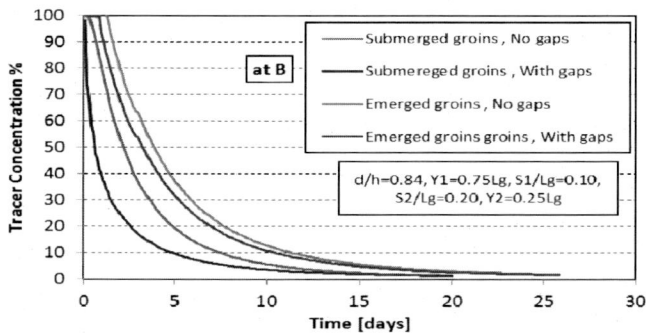

Figure 4.23: Time series for concentration of tracer material for various cases of groins height (emerged/submerged) and with/without gap at station (B).

Figure 4-24 shows the effect of the eastern gap location/width on the flushing process. It can be observed that the eastern gap location/width have significant effect on the flushing process. It can be seen that the flushing time ranges from 5-7.5 days in all cases if a gap exists in the eastern groin. The flushing time decreases with increasing the eastern gap width. Also, the flushing time decreases when the gap is located about 0.25Lg from the shoreline. This could be due to creating a short cut for the water particles to flow out of the perched beach. However, attention should be made that the gap cannot be constructed closer to the shoreline due to the aggressive erosion anticipated in the shoreline.

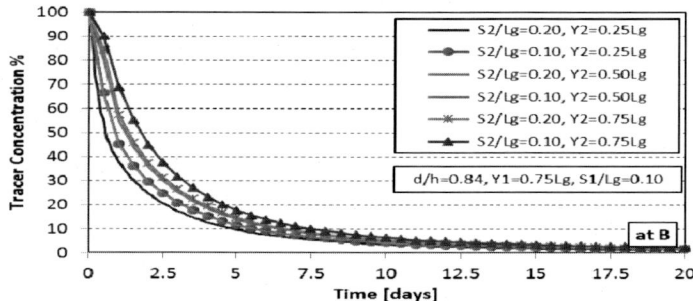

Figure 4.24: Effect of the eastern gap location/width on the flushing process.

4.4. Conclusions

A study has been conducted to investigate various configurations of the perched beach including submergence ratio of the breakwater, groin with/without gap, the gap width/location and emerged/submerged groins. The followings have been concluded:

- Large current velocities and eddies are generated offshore the submerged breakwater and off shore the west groin. The need for extra toe protection at these locations is far evident than other locations. Attention should be given to bed scour and toe design.
- The wave energy dissipation depends greatly on breakwater height. It is shown that the higher the breakwater is, the lower the transmitted but the slightly higher the reflected wave energies are.
- The higher the breakwater is, the lower offshore currents are. It should be taken into consideration to pay close attention to the breakwater toe and the swimming close to the breakwater due to large currents occurring around the breakwater.
- The effect of the submergence ratio (d/h) on the rip current is small compared to their effect on the long shore currents.
- The submergence ratio has significant effect on the shoreline changes and the accretion is generally slightly larger than erosion.
- There is no significant effect on the wave height at the centerline of the perched beach due to the gaps in groin and the emerged/submerged groins for the cases investigated.
- The emerged/submerged groins have significant effect on the cross and longshore currents inside the perched beach. On the other hand, the gaps in groin has no effect on the cross and longshore currents inside the perched beach for the studied cases.
- The height of the groins has significant effect on the shoreline changes along the up drift and down drift of the perched beach. Also, the emerged/submerged groins have significant effect on the shoreline changes along the up drift and down drift of the perched beach, as well.

- The accretion is slightly slower in case of the gap than in case of no gap. On the other hand, the down drift side has been found to erode at slow rate.
- Considerable shoreline changes have occurred along the shoreline in the vicinity of the perched beach. These changes diminish within three times the length of the groin on both sides of the perched beach.
- The accretion on the up-drift side decreases with increasing the western gap width. Also, erosion decreases on the down-drift side with increasing the western gap width. Moreover, accretion occurs inside the perched beach and increases as the western gap width increases.
- If the western gap width is $0.1Lg$ and constructed at 0.75 of the groin length very limited sedimentation is developed inside the perched beach, while larger gap width causes accelerated accretion inside the perched beach.
- Accretion inside the perched beach and erosion on the down drift decreases when the eastern gap is closer to the shore.
- There is no significant effect on the wave height along the centerline of the perched beach as the gap location/width varies for the cases investigated. Also, the gap location/width has no significant effect on the cross and longshore currents inside the perched beach for the cases investigated. However, the gap location/width has significant effect on the shoreline changes and the flushing.
- The submergence ratio has significant effect on the flushing time. The higher the breakwater is the higher the flushing time.
- Flushing process is significantly higher in case of submerged groins than in case of emerged groins. Also, flushing process is slightly higher in case of the gaps than in case of no gap. However, attention should be made that higher flushing is caused by higher waves and currents in the perched beach and this may affect the convenience of swimmers and may lead to possible dragging of swimmers out of the perched beach. A

compromise should though between the need for flushing and convenience/safety of swimmers in the perched beach.

- The flushing time decreases with increasing the eastern gap width. Also, the flushing time decreases when the gap is closer to the shore but it is not allowed to directly be at the shoreline.

Chapter 5
Study Area and Model Calibration

5.1. General

The present study has been applied on a part of the northern coastline of Egypt along the Mediterranean Sea. A chain of tourist villages and beautiful recreational beaches characterizes this part. This chapter describes the selected site information, the data collection technique. The bathymetric survey, bed material distribution, sediment transports and coast dynamics, general climate conditions and prevailing wave conditions were recognized. The model was calibrated to get the model empirical parameters and validated against the results acquired from the collected data and the field measurements.

5.2. Description of the Study Area

5.2.1. Location

Delft hydraulics divided the Egyptian northern coastline into number of cells based on the shape of the shore line and the main head lands (Delft Hydraulics Report (I), 2002).

The cell is defined as a unit in which the long shore morphological interaction is anticipated to be relatively strong. The cell is bounded by physical boundaries, e.g., rock head lands or harbors, such that interruption of long shore current takes place. The main coastal cells are shown in Figure 5.1 and Table 5.1.

The study area is located in cell (5). This area extends from El-Dikheila in the east to Ras El-Shaqiq in the west. The coast is sandy, locally backed by sand dunes and limestone ridges. This part of the coast is characterized by a chain of tourist villages and beautiful recreational beaches such as Marina village, Green, Marballa, Suez Canal village, Al-Hamam, Selsabil, Haydi, Ghernatah, Marakia, Burg Elarab and SidiKerir.

The pilot area is the coast of anew tourist resort located at SidiKrir kilometers 39.774/40.078 west of Alexandria along the Egyptian

Northwestern Mediterranean coast, as illustrated in Figure 5.2.

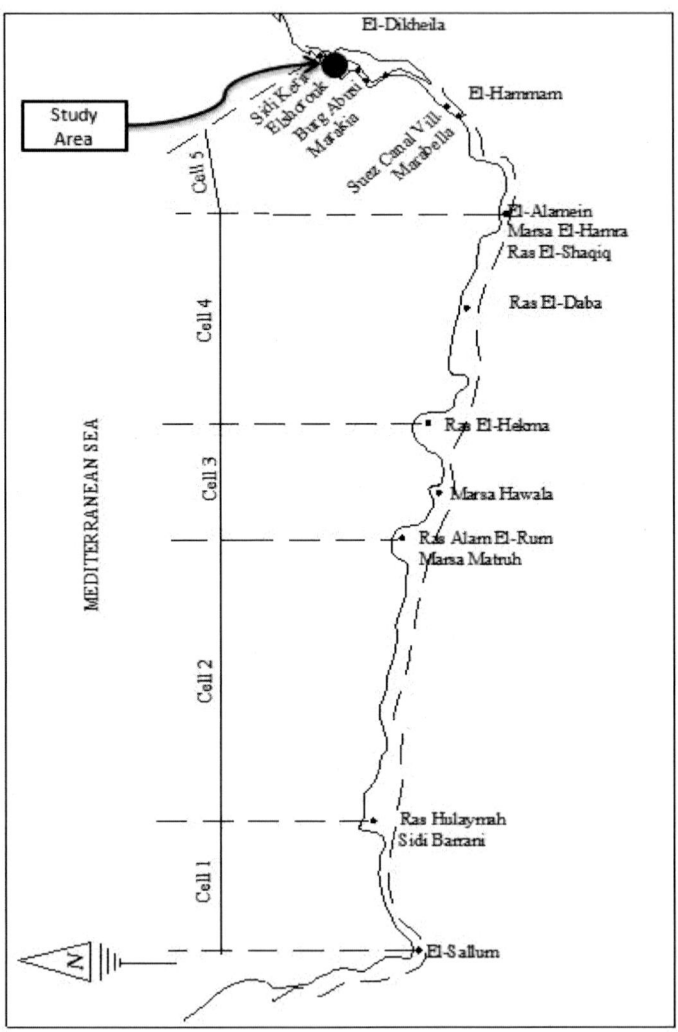

Figure 5.1: Study area and coastal cells (Delft Hydraulics Report (I), (2002)

Table 5.1: Coastal cells in the North-West coast of Egypt

Cell No.	Location	Main features	Length (approx.) (km)
1	El-Sallum to SidiBarrni	Sand interrupted by rock, backed by dunes and ridges.	90
2	SidiBarrni to Matruh	Alternating rocky and sandy stretches.	150
3	Matruh to Ras El-Hekma	Sandy bays between rocky outcrops.	70
4	Ras El-Hekma to Ras El-Shaqiq to (Al-Alamein)	Long alternating rocky and sandy stretches.	120
5	Ras El-Shaqiq to El-Dikheila	Sand beach backed by sand dunes.	120

The shoreline at the location of the new tourist resort makes $40°$ with the North direction and has a length of 300m, as shown in Figure 5.3. By analyzing the wave direction that attacks the coastal area, it could be concluded that the predominant wave is perpendicular to the shoreline approximately. So, the slope of the shoreline at the surf zone is steep due to the cross shore currents.

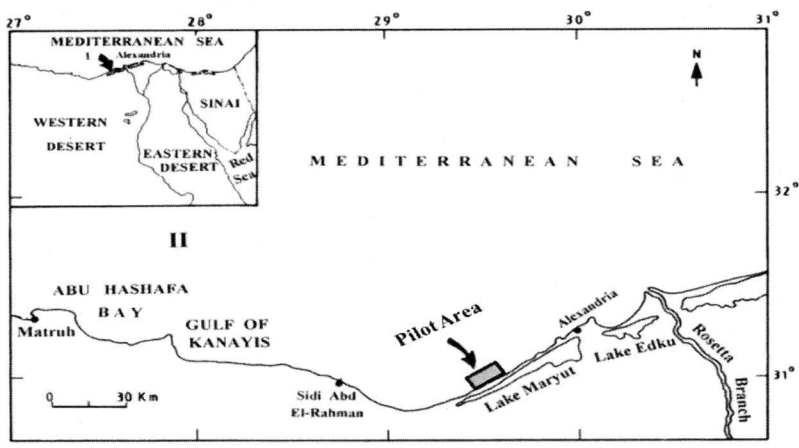

Figure 5.2: Location map of the pilot area

Figure 5.3: Satellite image of study area.

5.3. Data Collection and Analysis

Field measurements have been carried out to collect the following data:

- Bathymetric survey.
- Bed material distribution.

Another set of data regarding the waves, sediment transports, coast dynamics, general climate conditions, tide and currents were collected from the available sources such as scientific publication by members from Delft hydraulics, the Coastal Research Institute (CoRI) and the Hydraulics Research Institute (HRI). The following sections discuss the analysis of all collected data.

5.3.1. Bathymetry

Bathymetry survey is performed for the project site coastal area for an alongshore distance3000 meters; 300 meters the water front of the project site and 1350 meters from the two boundaries. The survey is performed from onshore the high water line to 10 meters and more offshore relative to Admiralty Chart Datum (A.C.D.).The shoreline is surveyed for about 2 kilometers on both sides of the project water front. The survey outputs illustrate that the shoreline is quasi-straight oriented at 40 degrees to the East of the North.

The survey was performed in May 2009using echo-sounder and differential GPS. Total station, stands with reflectors are used for the beach face and backshore. The water depths are corrected to account for sea water level fluctuations to be relative to A.C.D. Also, another bathymetry survey was carried out by the resort owner in December 2007, these data is used to calibrate the numerical model.

The data of the used equipment for bathymetric survey work are as the following:

- **Geographical Positioning System (GPS):**

It is used for measuring the global coordinates, it has been employed to record the boat position as well as the bed level (as an output of the echo sounder). The data acquisition has been done by the computer. The used GPS is GARMIN GPS 60 as shown in Figure 5.4.Its range is 0-100 km and its accuracy is ± 0.10m.

Figure 5.4: GPS adopted in the bathymetric survey (GARMIN GPS 60).

- **Echo Sounder:**

The used Echo Sounder is a compact and advanced precision echo-sounder. It was used for measuring the bottom profile. The data (the sea bed configurations) is recorded in the internal memory using a built-in recording system and then transferred to the computer. The used Echo Sounder is a GARMIN Echo 300C, as shown in Figure 5.5. Its range is 0-90 m in saltwater and its accuracy is ± 0.05 m.

- **Total Station (TS):**

The Total Station has been used for measuring the horizontal distances, horizontal and vertical angles. The data is recorded on a special tape using a built-in recording system. The data is easily transferred to the computer. The system is operated by a rechargeable battery of life for 8 hours. Its model is Topcon GTS-102N and its accuracy is ±0.003 m and 1.5 degree. Figure 5.6 shows the used Total Station during the survey works.

- **Rubber Boat:**

Fiber rubber boat was used in the bathymetric survey, as shown in Figure 5.7. Also, different boats were used with capacity up to 10 persons, and 55 HP outboard motors.

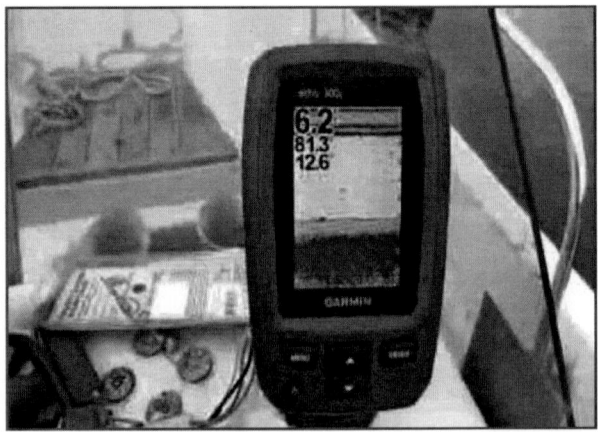

Figure 5.5: GARMIN GPSMAP Echo Sounder for recording bed levels.

Figure 5.6: Total Station used during the survey works of the shallow water depths.

Figure 5.7: Fiber rubber boat used in the bathymetric survey for water depths > 2m.

The bathymetry of the area, used in the modeling of waves and shoreline changes, has been based on the survey of SidiKrir region over an area of 3 km parallel to the shore line and 1.5 km offshore, as shown in Figure 5.8. The survey area has been conducted over a distance of 1500 m at the up drift of the village resort and 1500 m on the down drift side. From these hydrographic surveys, it has been found that the contour lines are rather straight and parallel to the shoreline. There are some offshore bars separated by gabs. The beach seem to be rather stable with minimum local shoreline changes. Many beach profiles have been drawn at a spacing of approximately 50m along the shoreline. The bed levels of each beach profile have been carried out by two methods:

- From the land side, the sea bed levels have been surveyed using reflector prisms and total station for the shallow water zone (a water depth of less than 2m).

- From the sea side, the sea bed levels have been measured by sounding using an Echo sounder installed on a rubber boat. The boat moves from the border of the shallow area till it reaches a water depth of 10 m.

The measured data have been used to draw the contour map for the surveyed area using SMS-10.1 model, as shown in Figure 5.5. It is noticed that all the shoreline features have been presented in the contour map. Bathymetric maps with UTM projection coordinates and contour step of 1.0 m have been presented. The survey showed that the nearshore profile of the coast is rather uniform, as demonstrated in Figure 5.9. The measured beach profiles are plotted along the project site water front as illustrated in Figures 4.10-a through 4.10-c.

Figure 5.8: Boundary of the bathymetric survey area.

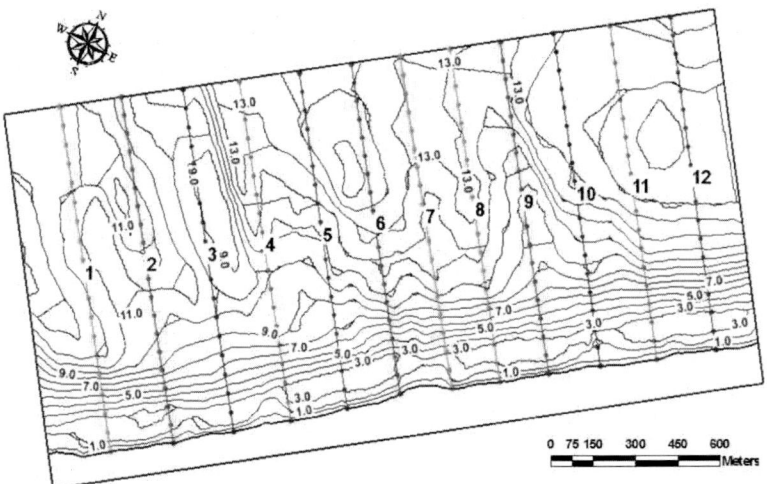

Figure 5.9: Bathymetric survey in May 2009and location of cross shore profiles.

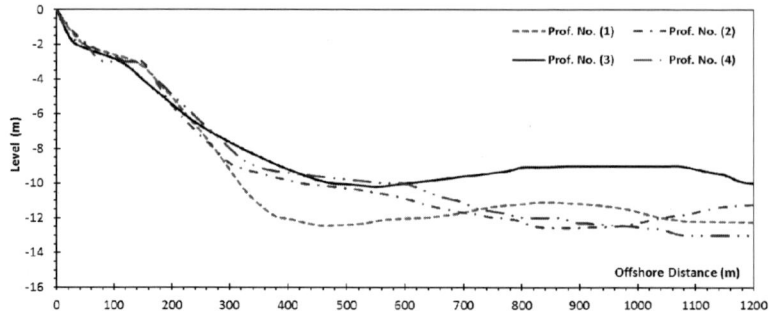

Figure 5.10-a:Cross shore profilesNo.1, 2, 3 and 4surveyed in May 2009.

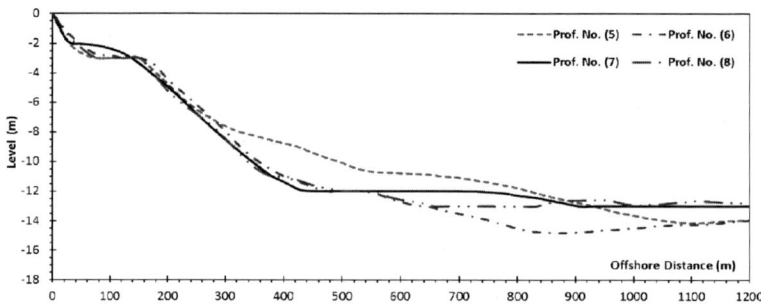

Figure 5.10-b:Cross shore profiles No.5, 6, 7 and 8 surveyed in May 2009.

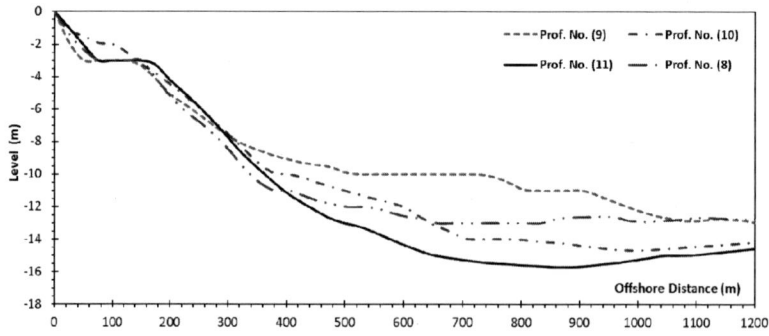

Figure 5.10-c: Cross shore profiles No.9, 10, 11 and 12 surveyed in May 2009.

5.3.2.   Bed Material Distribution

The sediments along the coast west of Alexandria consist mainly of calcareous material, which originates from the limestone rocks and ridges in the sea on the land in that area. These sediments differ considerably from those found east of Alexandria, which originate from the Nile delta.

Sediment samples at the study area region were collected at different depths/locations along the coast. Also, some near shore bore holes were implemented, as shown in Figure 5.11.

Figure 5.11: Bore holes and soil samples taken from the site in 2012.

Sea bed soil samples are collected from the backshore and the beach face up to 6 meters water depth. Laboratory grain sizes distribution tests performed for the samples illustrate that the backshore and the onshore part of the beach face consist of well graded medium sand. Samples collected in water depths close to 2 meters are medium to coarse sand which could be attributed to the washing of fine sand during collection of samples by divers.

The Sediment samples have been analyzed and the results are presented in Table 5.2. From analysis of the sediment sample in the study area, it is cleared that the sand on the beach and in the upper part of the coastal profile is coarse, in the order of D_{50}=(0.3-0.7) mm, while in the deeper part (below MSL-2m) the median grain size somewhat smaller, in order of D_{50}= (0.2-

0.3)mm.

Table 5.2: Sediment samples at different locations along the coast

Profile No.	Depth (m MSL)	D_{50} (mm)	Percentage of silt (%)
2	+1.0	0.70	0.01
	-2.0	0.26	0.02
	-4.0	0.24	0.02
	-6.0	0.19	0.03
4	+1.0	0.34	0.02
	-2.0	0.23	0.01
	-4.0	0.26	0.02
	-6.0	0.24	0.12
5	+1.0	0.58	0.04
	-2.0	0.18	0.02
	-4.0	0.25	0.03
	-6.0	0.24	0.05
6	+1.0	0.72	0.01
	-2.0	0.22	0.03
	-4.0	0.39	0.04
	-6.0	0.18	0.00
7	+1.0	0.33	0.02
	-2.0	0.23	0.02
	-4.0	0.26	0.02
	-6.0	0.55	0.12
8	+1.0	0.38	0.00
	-2.0	0.28	0.01
	-4.0	0.24	0.02
	-6.0	0.25	0.03
10	+1.0	0.55	0.03
	-2.0	0.18	0.03
	-4.0	0.26	0.02
	-6.0	0.24	0.04
12	+1.0	0.56	0.03
	-2.0	0.18	0.01
	-4.0	0.28	0.04
	-6.0	0.23	0.04

5.3.3. Sediment Transport and Coast Dynamics

The net long shore sediment transport in pilot area is small. On the basis of computations and observations near the south-western boundary of the pilot area, the net long shore transport is estimated to be slightly north-east directed with a magnitude of $15000m^3$/year approximately (HRI, 2004). Near the north-eastern boundary of the pilot area the net long shore transport is estimated to be small. During period of relatively high waves there will tendency for offshore directed sediment transport, resulting in some erosion of the upper part of the profile and deposition in the deeper part due to the steep profile. This effect is particularly pronounced during storm conditions. During low wave conditions there will be tendency for onshore directed transport during which the upper part of the profile will (partly) recover.

5.3.4. General Climate Conditions

The climate in the northern coastline of Egypt is characterized by a warm season (June to September) and cools one (November to April).The climate is mild in May and October most of time. Table 5.3 shows the climate condition in Alexandria (Alex.) and Port Said (P.S) of Egypt as measured and/or report by many researchers. Sharaf El-Din (1974), Sliahin (1985) and Hogbenet. Of. (1986), where f is the frequency in % and N, E, W, S, NW, NE, SE, SW are the directions. It is clear that the temperature of water varies from 16°c to about 26°c in Alexandria and it reaches a maximum of 27°c in Port Said. The highest rain fall occurs in December reaching 56.5 mm/month. The wind speed varies from 3 to 5 m/s along the coasts of Alexandria and Port Said. The average of the highest ten percent of the waves varies from 3.5m to 3.9m in winter, while it varies from 1.9 to 2.7m in summer.

Table 5.3: General climate conditions along the North coast of Egypt.

Parameter	J	F	M	A	M	J	J	A	S	O	N	D	Year	Unit	Ref.
Wind Speed	4.4	4.6	4.5	4.3	3.9	4.2	4.5	4.1	3.6	3.1	3.3	3.5	4.0	m/s	Sh
Wind direction	SW	N	NE	NE	NE	N	N	N	N	NE	NE	SW	NW	N°E	

		W				W	W	W							
Mean sea level		44	40	38	36	36	42	52	53	48	44	47	49	44.1	cm	SD
	Dir	1st quarter		2nd quarter		3rd quarter		4th quarter						f	
		f	H10	f	H10	f	H10	f	H10					in	
Waves	W	21	3.9	24	3.2	30	2.7	18	2.9					%	Ho
(open sea)	NW	16	3.9	19	3.0	32	2.7	21	2.7					H10	
	N	10	3.8	10	2.8	13	2.7	13	2.7					in	
	NE	9	3.6	8	2.8	6	2.0	10	2.7					m	

References: Sh = Sliahin, (1985): SD = Sharaf EL-Din, (1974): Ho = Hogbenatal,(1986)

5.3.5. Tide, Currents and SLR

Sea water surface levels pattern along the Egyptian Mediterranean coast is complex because different water level fluctuations mechanisms can occur simultaneously and have comparable magnitudes. It is not unusual to have pronounced astronomical tide, wave setup and wind set-up. There are at least 15 storms during the winter period some of them cause considerable wind set-up close to the shore.

The tides are mainly semi-diurnal and the astronomical tidal range is very small. It is about 0.2m for neap tides and0.5m for spring tide. Winds and other meteorological phenomena cause deviations from the astronomical tides.

Long term sea water surface measurements are available at Alexandria harbour since year1950, the corresponding sea water surface level pattern (BECOM et al., 1977)is as follow:

High high sea level (H.H.S.L.) = + 0.66 m.

Mean high sea level (M.H.S.L.) = + 0.13 m.

Mean sea level (M.S.L.) = 0.00 m.

Mean low sea level (M.L.S.L.) = - 0.13 m.

Low low sea level (L.L.S.L) = - 0.51m.

The Admiralty Chart Datum (A.C.D.) is lower by 34 centimeters

Estimates of local future Sea Level Rise (SLR)due to global warming effect by the year 2100 at Alexandria and Port Said were found to reach

37.9 and 44.2 cm, respectively (Frihyet al.,2010).

Currents along the Egyptian Mediterranean coast are generally variable in directions and magnitudes due to combined effects of the different driving forces for the currents; i.e. wind, hydraulic gradients, density variations, astronomical and coriolis effects, etc.. The predominant surface currents offshore the Egyptian Mediterranean coast are generally from0.25 to 0.5 knot (E.C.S.I., 1980). The current directions are generally variable, the prevailing direction is eastward. Near the shore littoral currents due to wave breaking are the major currents. The littoral currents depend on the nearshore wave conditions, sea bed bathymetry and shoreline orientation relative to waves direction.

Short term currents measurements had been performed at Marakia resort and El Dab'apower station project located about 25 and 110 kilometers to the west of the project site. Currents measurements were performed offshore Marakia village (E.C.S.I., 1980) at two locations in water depth close to 10 meters. "Aanderaa" type current meters were installed at 5 meters from the sea bed at the two locations during the period September to November1980. The measured currents at the two locations are generally similar with a mean values equal to about 0.20 and 0.40 meter per second during September/October and November, respectively. The maximum recorded currents were about 1.00 and 1.20 meters per second during September/October and November, respectively. The most probable heading of the currents was 245 degrees. It should be mentioned that these current measurements are highly over-estimated as the "Aanderaa" type current meter results are strongly affected by the existence of waves; i.e. wave orbital velocities are superposed on currents.

El Dab'a current measurements (Suez Canal Authority, 1987), performed during the period from June 1982to May 1984, in 10 and 25 meters water depth showed that the current directions are predominantly parallel to the coast and that wind induced currents are significant in water depths greater than about 10 meters. The current direction is about 43% directed to the east sector and 21% directed to the west sector and 36% in other directions.

The current speed is less than about 0.10 and 0.15 meter per second for 50 and 80 percent of the time, respectively. The maximum measured current speed was about 0.50 meter per second and occurs not more than 0.25 percent of the time.

Large scale wind-generated circulations occur, producing flow which is predominantly from west to east. Close to the coast the wave-induced currents dominate. Long shore wave-induced currents are generated by obliquely incident waves. When waves reach relatively shallow water depths they break and generate a current in the surf zone. The width and seaward depth varies with the wave height, but can be roughly indicated that the wave-dominated zone is MSL-8.0m.

The flow velocities generated by the waves depend mainly on the wave height and the wave angle relative to the coast. An increase in wave height and in wave angle results in an increase of the wave-induced current. Since the wave directions in the study area are predominantly from north-westerly directions, averaged over the year the eastward wave-induced currents dominate the westward currents in most of the study area. The above described currents are all long shore currents, with their main component more or less parallel to the shore. In the near shore area also cross-shore currents may be generated by the waves, like rip currents and undertow. These currents largely affect the swimming conditions.

5.3.6. Wave Data

The wave climate in the Mediterranean Sea is highly seasonal in nature and is strongly related to the large scale pressure systems whose limits overstep the boundary of the Mediterranean Sea. During the winter season from November to mid-April, storms occur regularly and are responsible for the generation of the highest waves from the west and north-west directions. During this period about 15 storms occur on the average per year each with a duration of about 3 to 4 days.

During the summer season from May to October, surface winds blowing over the Eastern Mediterranean generate swell from mainly the north-west. and north directions. Spring and autumn are transitional periods with limited wave energy. October and May are the calmest months in the year.

The accumulated wave data were collected from the available sources such as scientific publication by members from Delft hydraulics, the Coastal Research Institute (CoRI), the Hydraulics Research Institute (HRI) and El Dikheila Port Authority.

The wave data were measured during 2001-2005 with a wave gauge that belongs to El-Dekhila Port Station located at a water depth of 13 m. The wave data were statistically analyzed to obtain the representative wave conditions. The significant wave's conditions (all year) presented in Table 5.4, while the probability of wave occurrence is presented in Table 5.5 and Figure 5.12.Figure 5.13 shows the characteristic wave rose at water depth of 13 m where the prevailing direction is North-West which is common along the Egyptian Mediterranean Coast.

Table 5.4: The significant wave's conditions at a water depth of 13m.

Wave Condition	Hs (m)	T (s)	Direction (°N)	Duration (days/yr)
1	0.45	5	35	60.31
2	0.45	5	286	89.74
3	0.45	5	330	94.44
4	0.45	5	360	81.37
5	1.77	7	286	3.523
6	1.77	7	330	7.96
7	1.77	7	360	17.88
8	3.34	9	330	3.33
9	5.19	10	330	1.44

Table 5.5: The probability of wave occurrence at a water depth of 13m.

Significant wave Height (Hs)	Direction												Total
	15-45	45-75	75-105	105-135	135-165	165-195	195-225	225-255	255-285	285-315	315-345	345-360	
< 0.25	9.89	2.59	0.92	0.81	0.81	1.07	1.35	1.52	5	5.48	8.8	8.07	46.31

0.25-0.75	3.31	0	0	0	0	0	0	0	2.83	6.25	12.45	8.48	33.32
0.75-1.25	0.13	0	0	0	0	0	0	0	0.57	2.04	3.68	4.93	11.32
1.25-1.75	0	0	0	0	0	0	0	0	0.05	0.60	1.16	2.85	4.66
1.75-2.25	0	0	0	0	0	0	0	0	0	0.24	0.59	1.23	2.06
2.25-2.75	0	0	0	0	0	0	0	0	0	0.04	0.35	0.64	1.03
2.75-3.25	0	0	0	0	0	0	0	0	0	0	0.15	0.26	0.41
3.25-3.75	0	0	0	0	0	0	0	0	0	0	0.15	0.20	0.35
3.75-4.25	0	0	0	0	0	0	0	0	0	0	0.08	0.04	0.12
4.25-5.25	0	0	0	0	0	0	0	0	0	0	0.16	0.09	0.25
5.25-6.25	0	0	0	0	0	0	0	0	0	0	0.06	0.02	0.08
> 6.25	0	0	0	0	0	0	0	0	0	0	0.04	0.01	0.05
Total	13.33	2.59	0.92	0.8	0.81	1.07	1.35	1.52	8.45	14.65	27.67	26.82	99.99

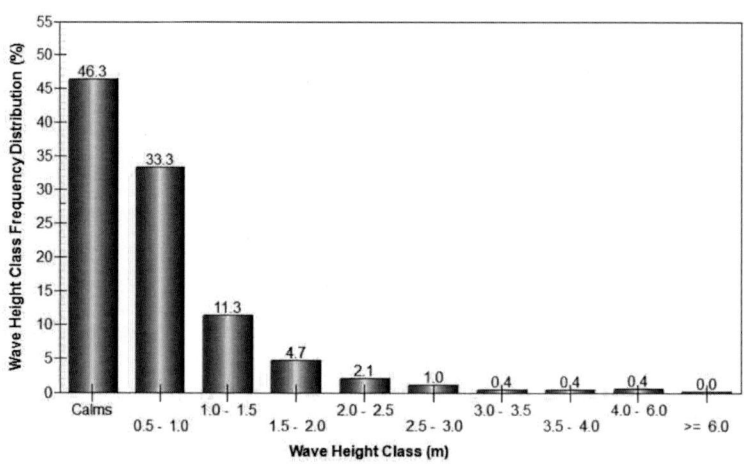

Figure 5.12: Probability of wave occurrence.

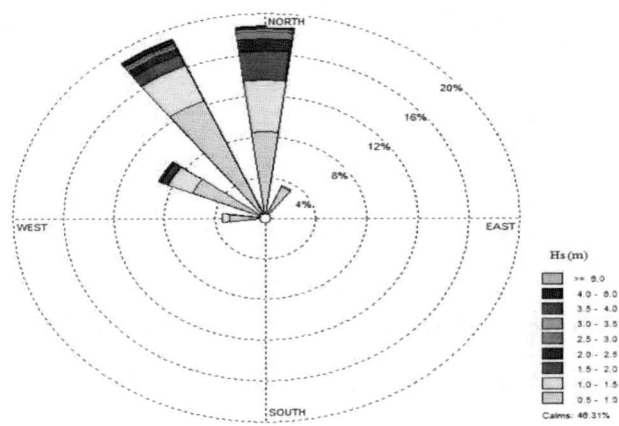

Figure 5.13: Significant wave height rose (all year)at a water depth of 13m.

The wave data has been analyzed and input to the numerical simulation model, namely as Surface-Water Modeling System (SMS) ver.10. SMS considers the existing shoreline as the baseline (initial shoreline) and the normal to it as a false North (direction=0 for the normal direction to the shoreline and angle is positive anti-clockwise). Figures 4.14-a to 5.14-c show the distribution of the wave height, wave period and wave direction during the period 2001-2005.It can be observed that in the summer season the waves from north-westerly directions are predominant while in the winter season direction are more variable with the highest waves approaching from north-westerly direction. Also, it can be observed that high waves occur for long period during the summer season, which is the most relevant for beach recreation and swimming. The wave period ranges from 5 to 10sec, as shown in Figure 5.14-c.

The spectral analysis of deepwater wave data during the period 2001-2005 is given in Table 5.6 as computed by SMS and presented graphically in Figures 4.15-a to 5.15-d. The wave angle is zero at the normal to the shoreline and positive anti-clockwise. It can be noticed that large numbers of small waves generally occur in the study area. The largest deep water wave being observed is about 5.42m and approaches the structure from its left hand side direction, i.e., at an angle of 18 degrees (see Table 5.6). Also, the average height of the highest 10% of deep water waves is as high as 2.5m.In the summer period the most relevant for beach recreation and swimming, wave heights only exceed a significant wave height (Hs) of 2.5m during (4-5) days, while a significant wave height of 1.77m is exceeded (20-30) days in the summer period and approaches the structure from its left hand side direction, i.e., at an angle of 44 degrees.

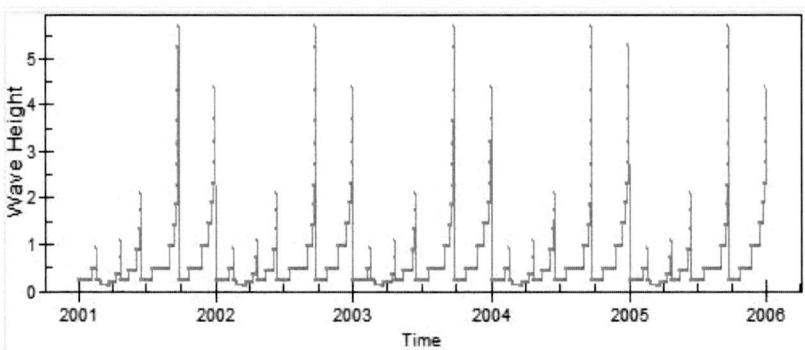

Figure 5.14-a: Wave height during 2001-2005.

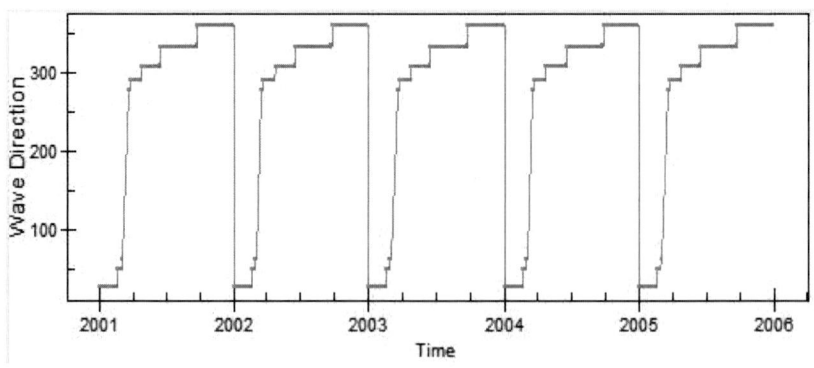

Figure 5.14-b: Wave direction from north during 2001-2005.

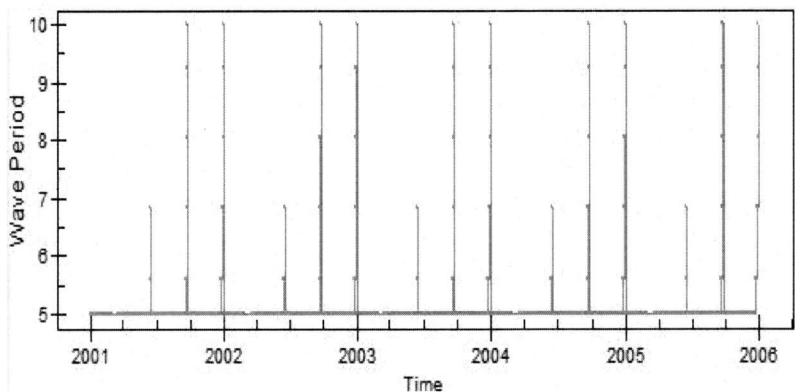

Figure 5.14-c: Wave period during 2001-2005.

Table 5.6: Spectral analysis of wave data in El-Dekhila Port Meteorological Station computed by SMS (2001-2005).

Index	Frequency of Occurrence	Hs (m)	Tp (sec)	Direction Angle (degrees)	Probability of Occurrence (%)
10101	405	0.14	5.0	-69	0.92
20101	6920	0.28	5.0	-38	15.79
30101	7254	0.36	5.0	-8	16.55
40101	9315	0.38	5.0	21	21.26
50101	5142	0.34	5.0	46	11.73
60101	4098	0.22	5.0	65	9.35
20102	55	0.93	5.0	-35	0.13
30102	2161	0.95	5.0	-8	4.93
40102	1611	0.95	5.0	21	3.68
50102	895	0.90	5.0	46	2.04
60102	250	0.72	5.0	63	0.57
30103	1250	1.43	5.0	-8	2.85

40103	510	1.42	5.0	21	1.16
50103	265	1.34	5.0	46	0.60
60103	20	1.08	5.0	63	0.05
30204	540	1.88	5.5	-7	1.23
40204	260	1.86	5.5	20	0.59
50204	105	1.77	5.5	44	0.24
30205	280	2.31	7.0	-7	0.64
40205	155	2.28	7.0	19	0.35
50205	20	2.09	7.0	42	0.05
30306	115	2.76	8.1	-7	0.26
40306	65	2.73	8.1	19	0.15
30407	90	3.22	9.3	-7	0.21
40407	65	3.18	9.3	18	0.15
30408	20	3.68	10.0	-7	0.05
40408	35	3.63	10.0	18	0.08
30409	36	4.37	10.0	-7	0.08
40409	70	4.31	10.0	18	0.16
30411	1	5.29	10.0	-7	0.00
40411	45	5.42	10.0	18	0.10
Calm events	1771	--	--	--	4.04
Total	43824				100

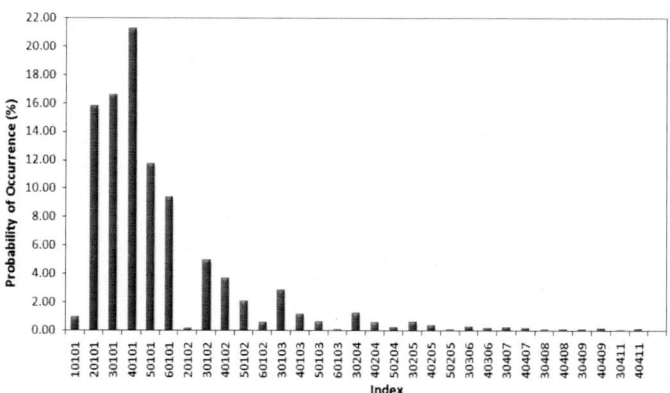

Figure 5.15-a: Graphical presentation of spectral analysis of wave data in El-Dekhila Port Meteorological Station computed by SMS (2001-2005).

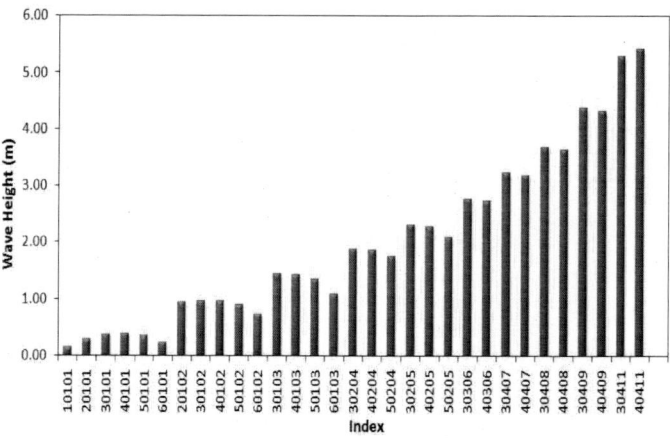

Figure 5.15-b:Graphical presentation of spectral analysis of wave height in El-Dekhila Port Meteorological Station computed by SMS (2001-2005).

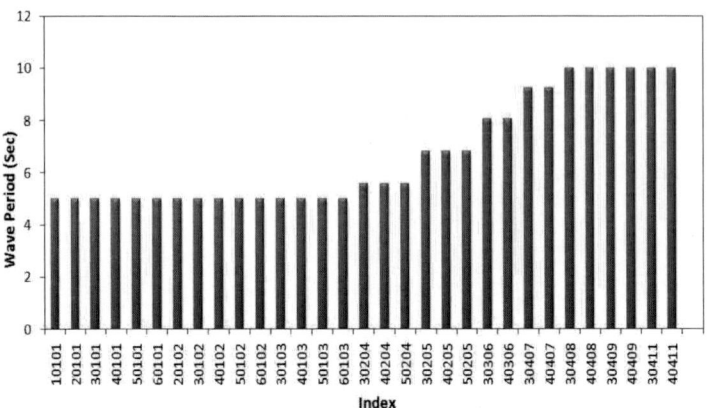

Figure 5.15-c: Graphical presentation of spectral analysis of wave period in El-Dekhila Port Meteorological Station computed by SMS (2001-2005).

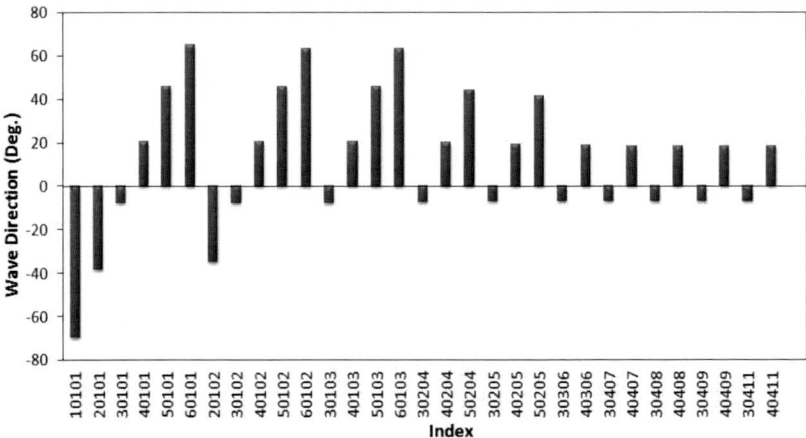

Figure 5.15-d: Graphical presentation of spectral analysis of wave direction in El-Dekhila Port Meteorological Station computed by SMS (2001-2005).

The distribution of the deepwater wave energy (m²/Hz) versus the direction angle measured from the normal to the shore line during the occurrence of maximum incident wave height ($H_{max.}= 4.5m$) is shown in Figure 5.16-a. It can be observed that most of the wave energy is centered about the normal to the shoreline and the maximum is reached at about 18 degrees to the normal on the shoreline. Also, the distribution of the deepwater wave energy versus the frequency and the spectral energy distributions during the occurrence of maximum incident wave height are shown in Figures 4.16-b and 5.16-c, respectively.

While, the distribution of the deepwater wave energy versus the direction angle measured from the normal to the shore line during the occurrence of the significant wave height ($H_s=2.5m$) is shown in Figure 5.17-a. It can be observed that most of the wave energy is centered about the normal to the shoreline and the maximum is reached at about 25 degrees to the normal on the shoreline. Also, the distribution of the deepwater wave energy versus the frequency and the spectral energy distributions during the occurrence of maximum incident wave height are shown in Figures 4.17-b and 5.17-c, respectively.

Moreover, the distribution of the deepwater wave energy versus the direction angle measured from the normal to the shore line during the occurrence of the prevailing wave height in the summer ($H_s=1.77m$) is shown in Figure 5.18-a. It can be observed that most of the wave energy is shifted to the left of the normal to the shoreline and the maximum is reached at about 44 degrees to the normal on the shoreline. Also, the distribution of the deepwater wave energy versus the frequency and the spectral energy distributions during the occurrence of maximum incident wave height are shown in Figures 4.18-b and 5.18-c, respectively.

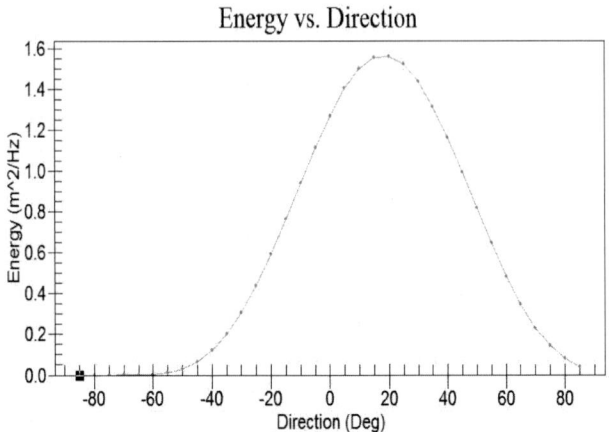

Figure 5.16-a: Wave energy distribution versus the direction angle
($H_{max.}= 4.5m$).

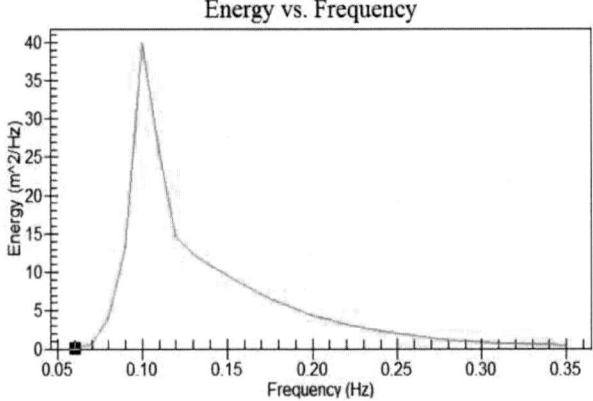

Figure 5.16-b: Wave energy distribution versus frequency ($H_{max.}= 4.5m$).

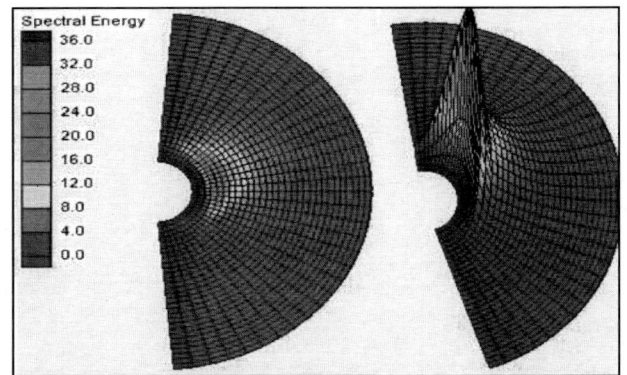

Figure 5.16-c: Spectral energy distributions ($H_{max.}$= 4.5m).

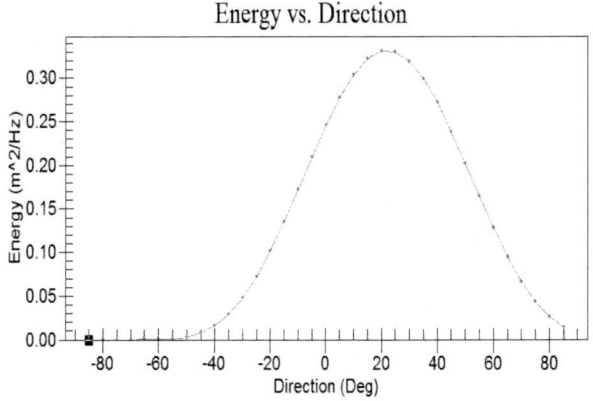

Figure 5.17-a: Wave energy distribution versus the direction angle (Hs = 2.50m).

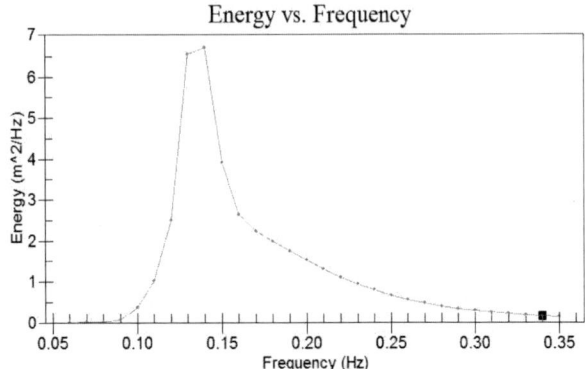

Figure 5.17-b: Wave energy distribution versus frequency (Hs = 2.50m).

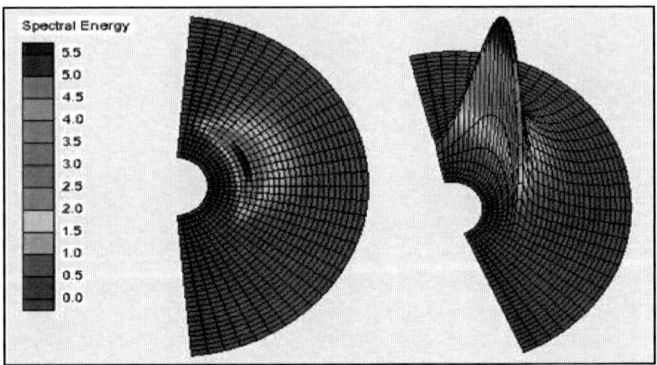

Figure 5.17-c: Spectral energy distributions (Hs = 2.50m).

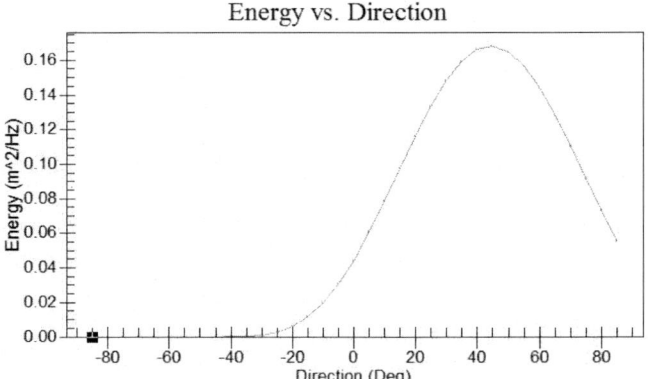

Figure 5.18-a: Wave energy distribution versus the direction angle (Hs = 1.77m).

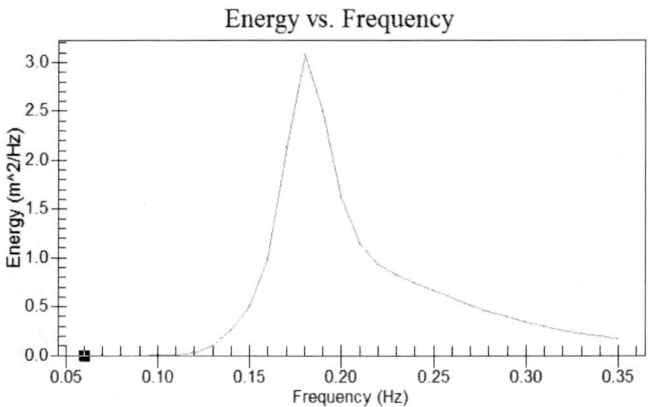

Figure 5.18-b: Wave energy distribution versus frequency (Hs = 1.77m).

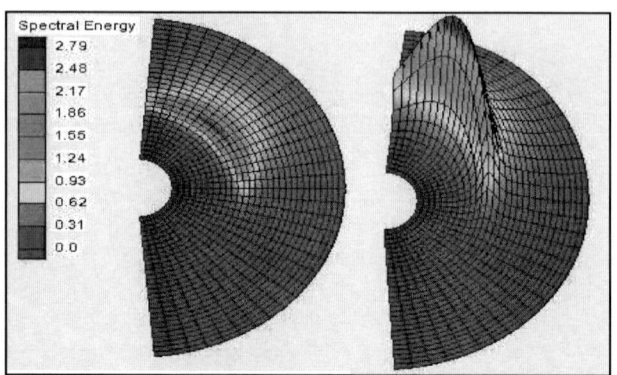

Figure 5.18-c: Spectral energy distributions (Hs = 1.77m).

5.4. Model Setup

5.4.1. Grid Description

Based on the bathymetric survey, rectilinear grids were generated and the depth files were prepared. For the wave models (CMS-WAVE&BOUSS-2D) a rectilinear grids were generated and the associated depth files were prepared. A high grid resolution was applied in the area of the proposed structure (perched beach) in front of the tourist resort, and a low resolution was implemented far away of the area of their influence. The grid cell size (5 x 5 m) was fine inside the structure area and the adjacent regions while it was course (20 x 20 m) in the far field till the open boundary. The model grid meets criteria for smoothness (adjacent cells do not differ much in size) and orthogonally (the angles of the comers of the cells are close to 90°), in order to avoid that small disturbances due to irregularities in the grid grow to governing features during the computation.

The computational grid for the hydrodynamic model is rectilinear and covers the area in the vicinity of the structure. Figure 5.19 shows an overview of the generated rectilinear grid. The model extends 3000 m along the shoreline and 1500m offshore. The model grid was generated using the software tool SMS-Cartesian grid module and its general characteristics are described as follows:

Model dimensions	Distance along the shoreline direction = 3000 m
	Distance perpendicular on the shoreline direction = 1500 m
Grid size	XxY = 240 x 150 = 36000 nodes
Cell length	min= 5m
	max= 20m
Cell width	min= 5m
	max= 20m

5.4.2. Model Topography

The bathymetric data from field measurements was used to generate the model topography. In an area of 500 m from the shoreline and at the area of the proposed structure the bathymetry was described at a high resolution. The second set has a lower resolution and describes the bathymetry for the areas far from the proposed structure. For a better hydrodynamic modeling for the area near the proposed structure, the total area of the model was extended to cover a 3.0 km along the shoreline. An over view of the topography of the simulated area with the Location of the proposed structure is given in Figure 5.20.

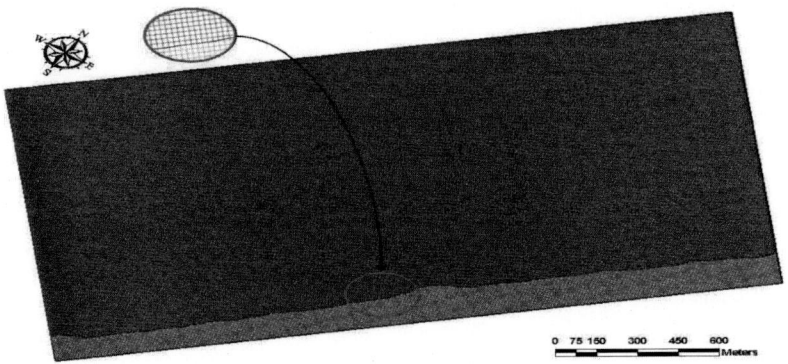

Figure 5.19: Grid of the simulated area.

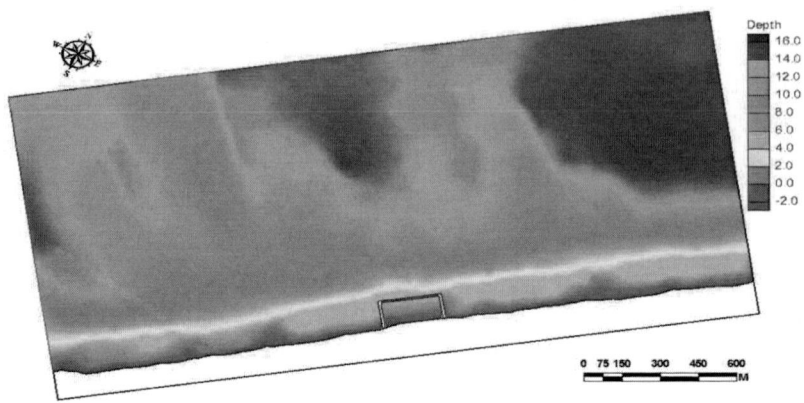

Figure 5.20: Model topography with the location of the proposed structure.

5.5. Model Calibration

Based on the measured shorelines of 2007, 2009 and 2012the model was calibrated. A comparison was prepared between the computed and measured shoreline changes (Figure 5.21) for different cases of different calibration coefficients (K1, K2) in the longshore sediment transport formula. Figure 5.22 illustrate the correlation between the measured shoreline and the computed shoreline for various values ofcalibration coefficients. The correlation value (R^2 =0.965) for K1=0.25 and K2=0.03. The correlation value (R^2 =0.9935) for K1=0.35 and K2=0.06. However, the correlation value (R^2 =0.9999) for K1=0.45 and K2=0.10. Thus, the best values of the calibration coefficients for the study areaareeK1=0.45 and K2=0.10.

As mentioned before,the net long shore transport is estimated to be slightly north-east directedwith a magnitude of 15000m³/yearapproximately (HRI, 2003).The longshore sediment transport were calculated based on Ozasa and Brampton formula in SMS model, and calibrated based on these amount of sediment transport. The computed net transport integrated over one year of simulation, including seabed update. The model computes the north-eastward directed net longshore transport of 5,000 to 12,000 m³/year at the updrift boundary. The overall conclusion of the confrontation of the numerical model results with the measurements is that the numerical model is capable of simulating the shoreline changes in the pilot area

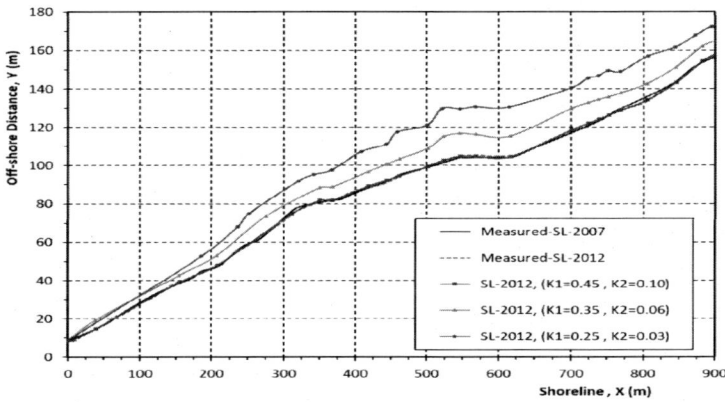

Figure 5.21: Shoreline change rates along the coast in term of the calibration coefficients.

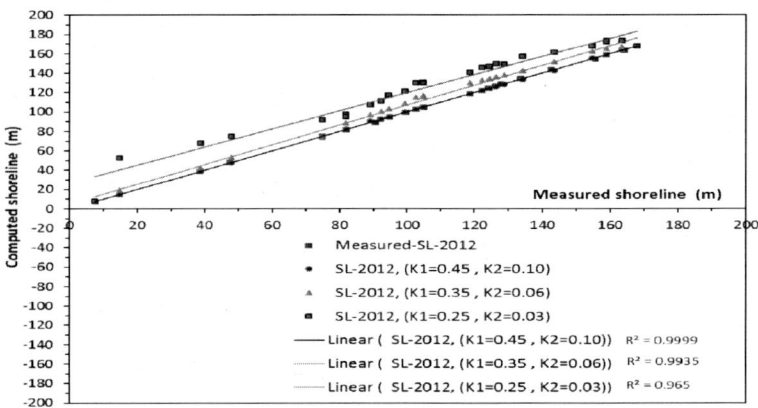

Figure 5.22: Correlation of the shoreline change for different cases of different calibration coefficients (K1, K2).

Chapter 6
Application to the North-West Coast of Egypt
6.1. General

A study has been made on the use of perched beaches to provide safe swimming conditions using SMS-10.1model for the simulation of wave kinematics and shoreline changes. Application has been made to an actual scale perched beach recently approved by EEAA and SPA. The perched beach was completed by 2013 and its data have been employed in the current work. Additional investigations have been conducted to extend the results and lessons learned from the existing design. Several alternatives have been examined and compared to develop general guidelines for similar beaches along the North-West coast of Alexandria, denoted as Cell 5 (see SPA 2002). Special attention has been given to alleviate the possible adverse impact of the protection structures. These impacts include excessive shoreline erosion and possible deterioration of water quality.

Field measurements were carried out in May 2009to prepare the required studies for the approval of the project. Another campaign of field measurements has been carried out before commencement of construction in 2011 and finally in 2013 after complete construction of the perched beach. Field measurements have been used in this study to calibrate the numerical model, i.e., SMS10.1. In this chapter, various configurations of the perched beach have been described, simulated and the results analyzed to provide general guidelines for the design of a perched beach within the Al-Arab Bay zone denoted as Cell 5 of the North West coast of Egypt. The proposed alternatives have various configurations including: submergence ratio of the breakwater, groin with/without gap and emerged/submerged groin. The alternatives have been compared from the point of view of wave height, currents velocities, flushing and shoreline changes.

6.2. Model Input

The layout and dimensions of the proposed perched beach is shown in Figure 6.1. The resort beach is 300m long, but the protected length is 200m leaving 50m on both sides of the groins open to the sea. However, simulations using SMS model have been made for 3000m along the shore line and extends for about 1500m normal to the shoreline to limit the boundary effects on the study area.The perched beach generally has two bounding groins and shore parallel breakwater at water depth equal to 3.5m. The groins extend for 120m normal to the shoreline and have some opening in its body, either pipe or a gap. The average effective grain size of the soil on-site is found to be 0.3mm, the average berm height is computed and found to be 2m and the closure depth is considered as 8m.

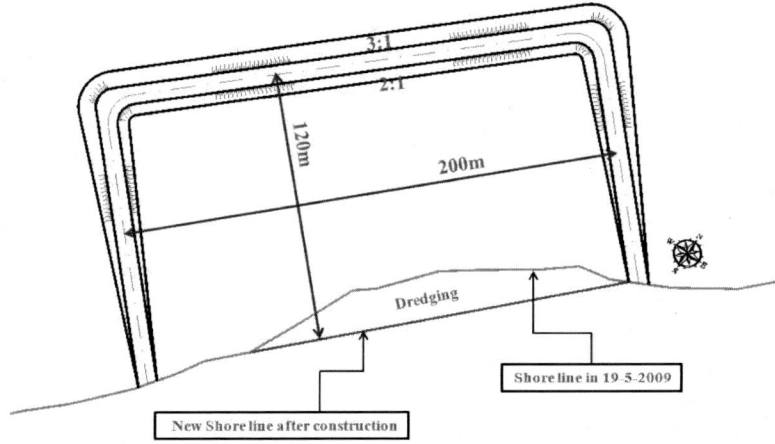

Figure 6.1: Layout of the proposed structure.

Simulations have been made for the case of before construction of the perched beach to evaluate the problems encountered in the beach. This case was called the baseline condition and shall be used asa reference condition. The proposed perched beach conceptwas first introduced and investigated numerically and experimentally by Delft Hydraulics, Netherlands (SPA, 2002) to develop safe swimming conditions.

Based on the results of the wave rose in El-Dekhiela port, it has been found that three wave conditions are of major importance, i.e., two during the summer season and one in winter, as shown in Table6.1.The dominant wave height all over the year in deep water is 1.77m and it occurs for 20-30 days during the summer season. However, in the summer period the most relevant for beach recreation and swimming condition, wave heights reach a significant wave height (Hs) of 2.5mfor (4-5) days.

Table 6.1: Dominate deep wave conditions.

Wave Condition	Hs (m)	T (s)	Duration (days)	Notes
1	1.77	5.0	20-30	In summer
2	2.50	7.5	4-5	In summer
3	4.50	10.0	0.86	In winter

The wave data and bathymetric survey have been input to run the model for alternative configurations, as presented in Table 6.2. The results of each alternative have been presented, analyzed, compared and recommendations are given.

Table 6.2: Summary of runs simulated by the SMS-10 model.

Run No.	Alternative No.	Wave Conditions		Breakwater	Groins	
		Hs (m)	T (sec)	Crest Level	Crest Level	Gap Width (m)
1	Base line	1.77	5.0	--	--	--
2		2.50	7.5	--	--	--
3		4.50	10.0	--	--	--
4	Alternative No. (1)	1.77	5.0	-0.50	-0.50	--
5		2.50	7.5	-0.50	-0.50	--
6		4.50	10.0	-0.50	-0.50	--
7	Alternative No. (2)	1.77	5.0	-0.50	+2.0	--
8		2.50	7.5	-0.50	+2.0	--
9		4.50	10.0	-0.50	+2.0	--
10	Alternative No. (3)	1.77	5.0	-0.50	+2.0	10
11		2.50	7.5	-0.50	+2.0	10
12		4.50	10.0	-0.50	+2.0	10

13		1.77	5.0	-0.50	-0.50	10
14	Alternative No. (4)	2.50	7.5	-0.50	-0.50	10
15		4.50	10.0	-0.50	-0.50	10
16		1.77	5.0	-1.25	-0.50	--
17	Alternative No. (5)	2.50	7.5	-1.25	-0.50	--
18		4.50	10.0	-1.25	-0.50	--
19		1.77	5.0	-0.90	-0.50	--
20	Alternative No. (6)	2.50	7.5	-0.90	-0.50	--
21		4.50	10.0	-0.90	-0.50	--

6.2.1. Alternative (1): Submerged Groins and Breakwaters

This alternative has shore parallel breakwater and two groins having their crest at -0.50m MSL. Figure 6.2 shows the layout of the structures and the bed levels as presented by the SMS model. This alternative has all the merits of submerged structures, e.g., aesthetic point of view, moderate waves, good flushing…

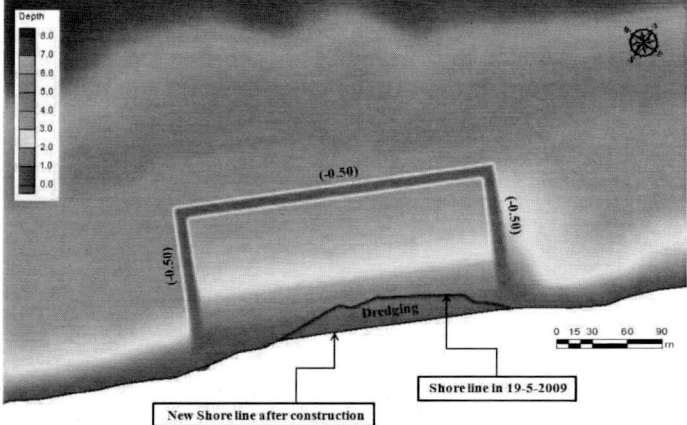

Figure 6.2: Layout of perched beach and sea bed levels for alternative (1).

6.2.2. Alternative (2): Submerged Breakwater and Emerged Groins

This alternative is similar to alternatives 1 except that the groins are constructed higher than the water surface at +2.0m MSL. Figure 6.3 shows the layout of structures and the sea bed levels as presented by the SMS model. This alternative enables the possible economic use of the groins for recreation activities, but it affects the sight distance of beach visitors to some extent due interruption of the sight line at the boundaries.

6.2.3. Alternative (3): Submerged Breakwater and Emerged Groins with a Gap

This alternative is similar to alternative (2), but the west groin has a gap located at 1/4 of the groin length measured from offshore head and the east groin has another gap at 1/4 of the groin length measured from the shoreline. Figure 6.4shows the layout of this alternative. The proposed gap are sought to help flush the protected area and allow less trapping of long shore sediments.

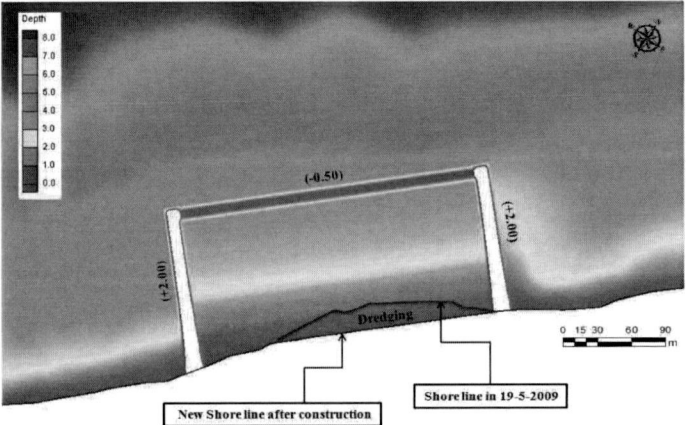

Figure 6.3: Layout of structures and sea bed levels for alternative (2).

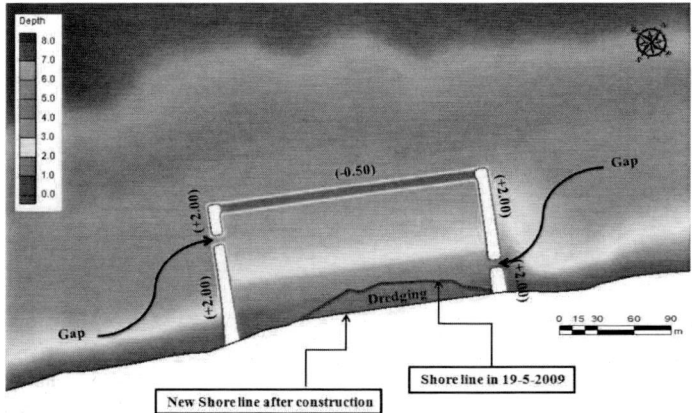

Figure 6.4: Layout of structures and sea bed levels for alternative (3).

6.2.4. Alternative (4): Submerged Breakwater and Groins with a Gap

This alternative is similar to alternative (3), but the crest level of the two groins is at -0.50m MSL. Figure 6.5shows the layout of this alternative and the topography of the sea bed generated by SMS model.

6.2.5. Alternative (5): Very Low Crested Breakwater and Groins

This alternative is similar to alternative (1), but the crest level of the two groins and the submerged breakwater is at -1.25m MSL. Figure 6.6shows the layout of this alternative and the sea bed levels generated by SMS model.

6.2.6. Alternative (6): Medium Low Crested Breakwater and Groins

This alternative is similar to alternative (1), but the crest level of the two groins and the submerged breakwater is at -0.9m MSL. Figure 6.7shows the layout of this alternative and the topography of the sea bed levels generated by SMS model.

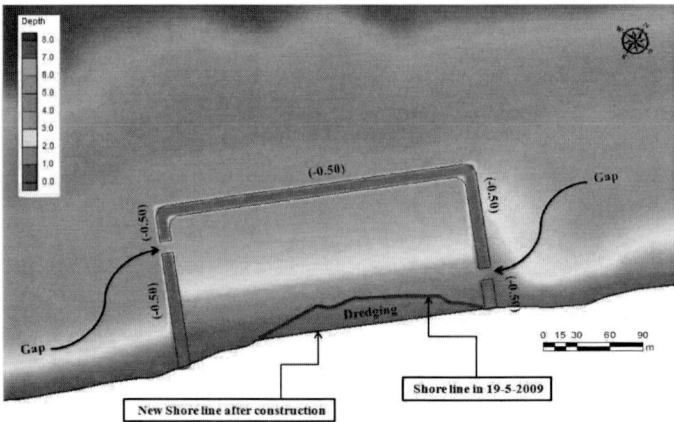

Figure 6.5: Layout of structures and sea bed levels for alternative (4).

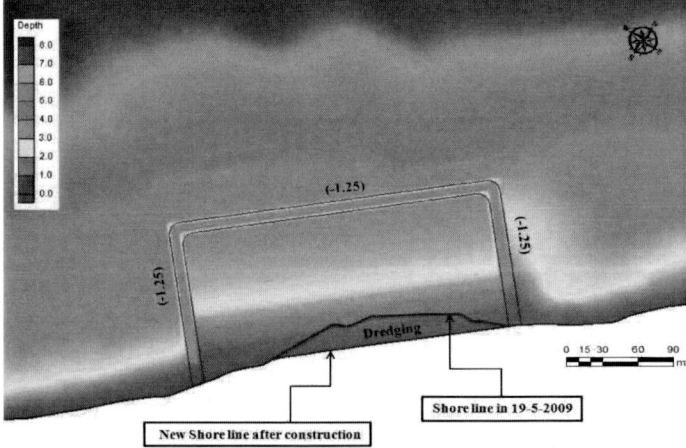

Figure 6.6: Layout of structures and sea bed levels for alternative (5).

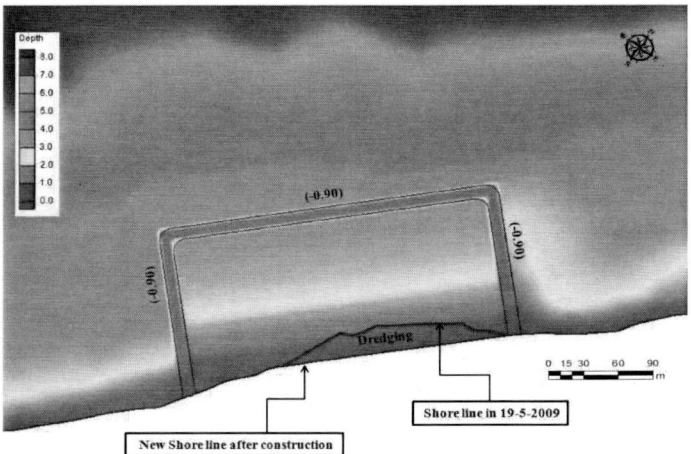

Figure 6.7: Layout of structures and sea bed levels for alternative (6).

The wave data and bathymetric survey have been used to run the model for the current configurations, as presented in Table 6.2.

In this section the results for each alternative including the baseline condition are presented, analyzed and discussed. The baseline and each alternative have been tested with three wave conditions; 1.77m; 2.5m and 4.5m approaching from the North West direction and represent moderate and sever wave conditions. The presented results are mainly focus on the flow field inside the perched beach and the adjacent area along the shoreline. Moreover, the shoreline changes (erosion/accretion) have been investigated.

6.3. Base Line Condition

Figures 6.8-a to 6.8-c show the wave direction and height as waves approach the shoreline for moderate and severe wave conditions. It can be observed that waves approach the shore normal to it due to refraction. The variation of wave heights along the centerline of the project area for the case before constructing the perched beach (Shoreline is located at distance = 0) is presented in Figure 6.9. It can be observed that the waves break at distance of 160, 120 and 40m measured from the shoreline for deepwater wave heights of 4.5, 2.5 and 1.77m, respectively. Also, it can be observed that only a narrow strip of less than 30m having a wave height of less than 0.60m for H=1.77m exist. Thus, the beach is unsuitable for swimming even during the occurrence of moderate wave heights. Also, it can be observed that for wave height of 4.50m, the wave height decreases rapidly when compared with incident waves of 1.77m and 2.50m height. Figures 6.10-a to 6.10-b show the radiation stress in the near shore zone. The location of the breaker line for the dominant high waves during the summer season, i.e., H=1.77m, is very close to the shoreline and hence limiting the surf zone against the convenience of swimmers. It can be observed that there are two breaker lines, i.e., in deep and shallow water. Moreover, strong rip current can be noticed on a rip head, which is formed offshore. Although the breaker line is shifted offshore and the surf zone becomes wider than the case of H=1.77m, considerably large waves exist near the shoreline and a secondary breaker line is formed. It can be concluded that swimmers should abandon the sea for long period during summer season due to the occurrence of H = 1.77 and 2.5m

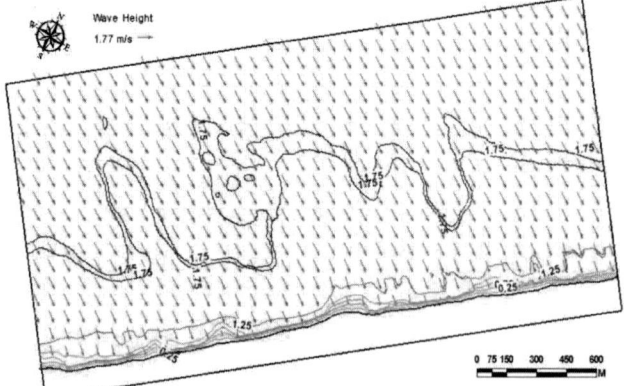

Figure 6.8-a:Computed wave height and direction in the study area (H=1.77m, T=5sec and direction is NW) for the baseline case

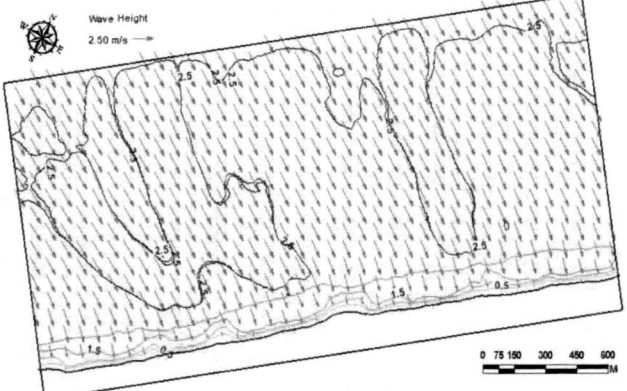

Figure 6.8-b: Computed wave height and direction in the study area (H=2.5m, T=7.5sec and direction is NW) for the baseline case

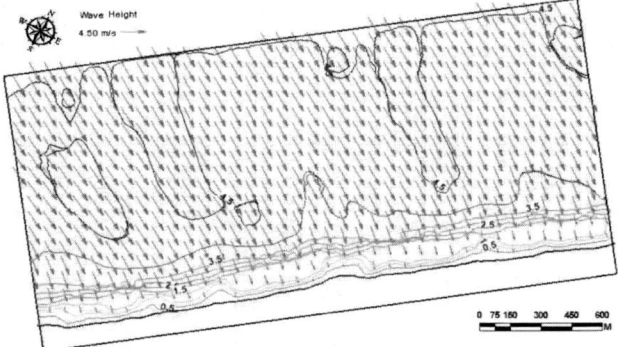

Figure 6.8-c: Computed wave height and direction in the study area (H=4.5m, T=10sec and direction is NW) for the baseline case

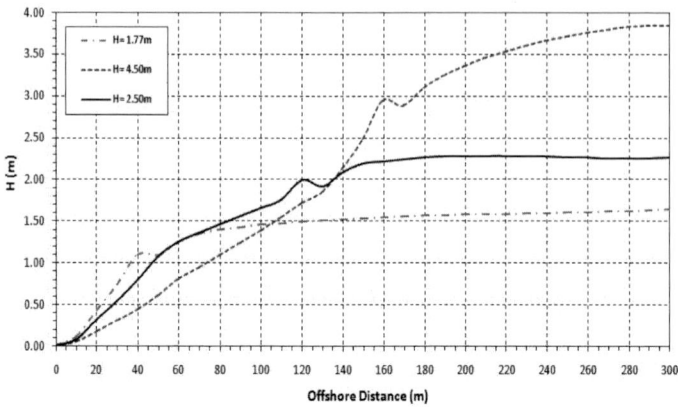

Figure 6.9: Computed wave height along the centerline of the beach.

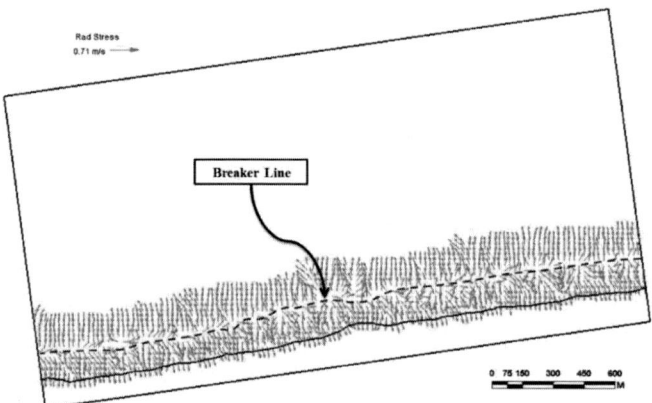

Figure 6.10-a: Computed radiation stress in the near shore zone and the location of the breaker line for H=1.77m.

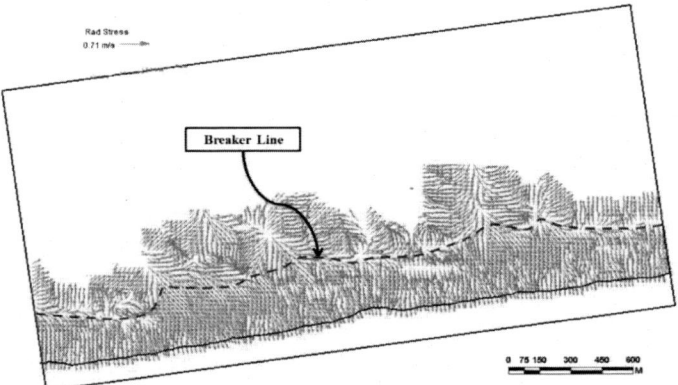

Figure 6.10-b: Computed radiation stress in the near shore zone and the location of the breaker line for wave heights 2.50m and no structures.

Figures6.11-a to 6.11-c show the flow field in the pilot area. Five transects along the shoreline of the resort are shown in Figures 6.12-a & 6.12-b, namely as C.S.1~C.S.5 and C.S.3 is at the resort centerline. Furthermore, the cross shore and the long shore currents are also computed as shown in Figures 6.13-a to 6.13-f.

It is clear that the transect C.S.3 shows the largest long shore velocity and has a maximum of 0.55m/s for H=1.77m and 0.74 m/s for H=2.5m and H=4.5m. On the other hand, the rip current reaches a maximum value of about 0.52m/sec at C.S.3 for H=1.77m, but a maximum value of 0.75 m/s occurs at C.S.4 for H=2.5m during the summer season. Furthermore, a maximum value of 0.77 m/s occurs at C.S.2 for H=4.5m. It can be confirmed that the rip currents are generally large and dangerous for more than a month every year during the summer season (see Table 6.1).

It can be concluded that the presence of the hazardous rip currents, circulation zones and the adverse effect of the steep profile along the coast of the pilot area in combination with the breaking waves makes that only a very narrow strip (20m approx.) of coastline usable by most swimmers. So, there is an essential need to construct suitable coastal structure in the pilot area in order to create safe conditions for swimmers with minimum shoreline change.

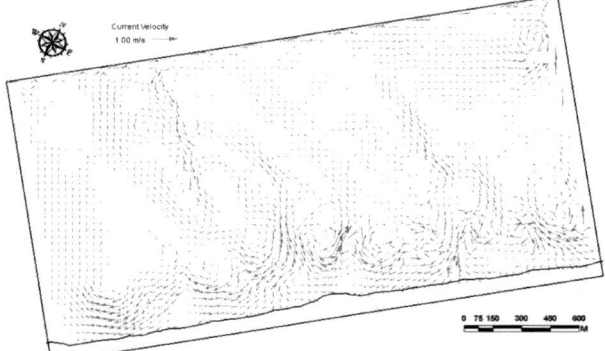

Figure 6.11-a: Current velocity in the near shore zone with wave 1.77m.

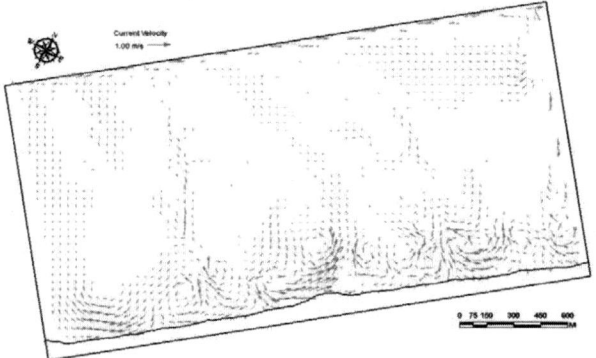

Figure 6.11-b: Current velocity in the near shore zone with wave 2.50m.

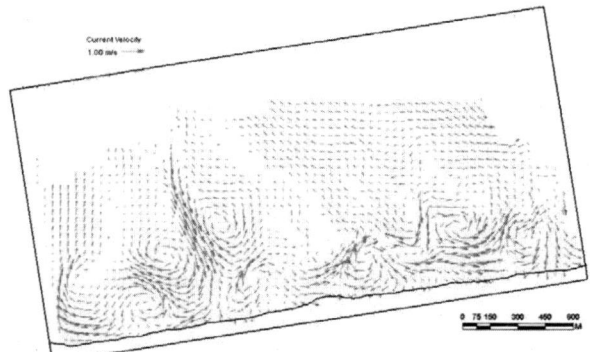

Figure 6.11-c:Current velocity in the near shore zone with wave 4.50m.

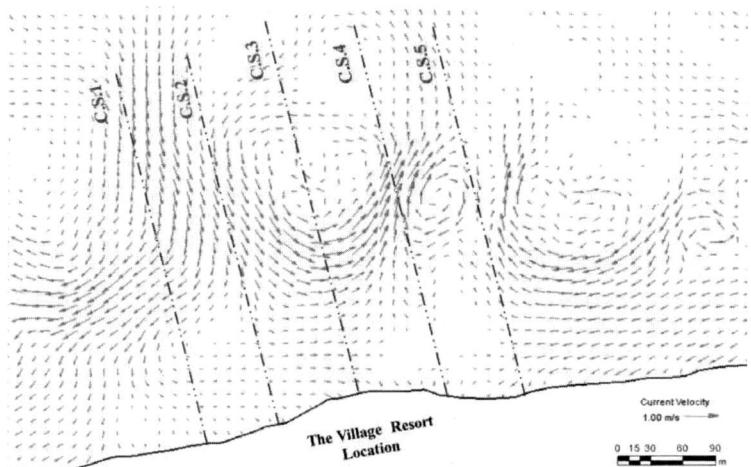

Figure 6.12-a: Current velocity in the beach for H=1.77m, T=5s and direction is NW.

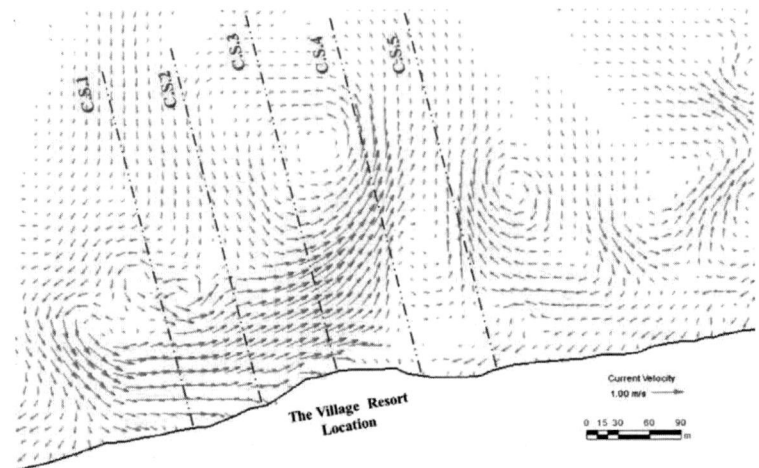

Figure 6.12-b: Current velocity in the beach for H=2.5m, T=7.5s and direction is NW.

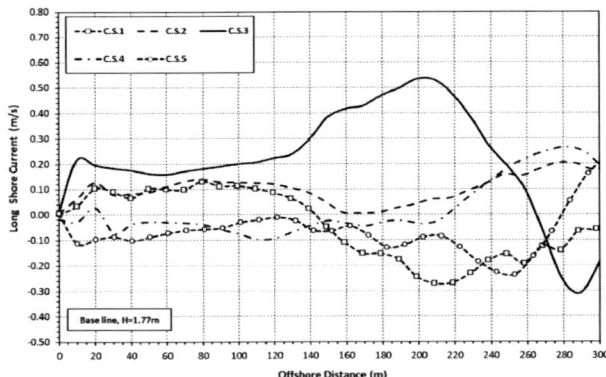

Figure 6.13-a: Long shore current at the profiles for H=1.77m.

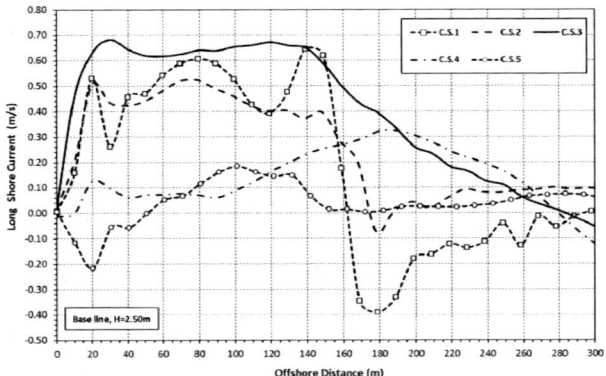

Figure 6.13-b: Long shore current at the profiles for H=2.50m.

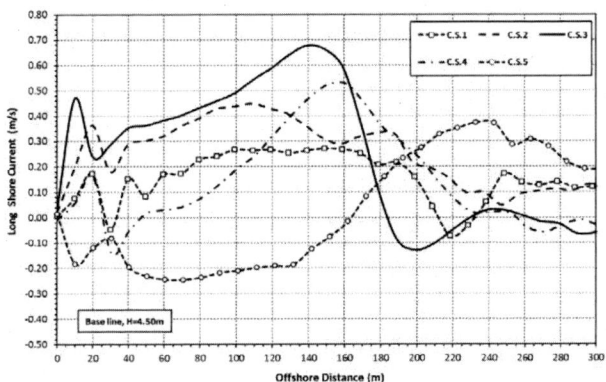

Figure 6.13-c: Long shore current at the profiles for H=4.50m.

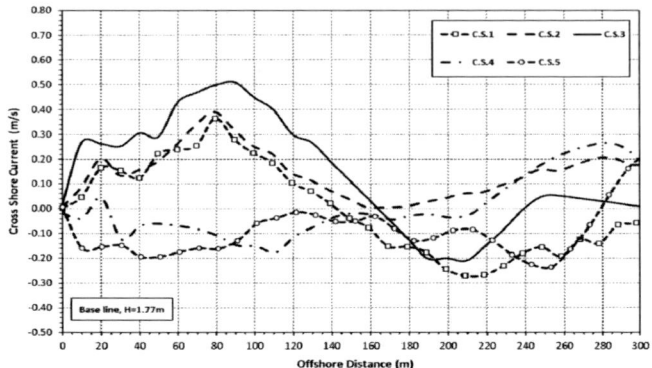

Figure 6.13-d: Cross shore current at the profiles for H=1.77m.

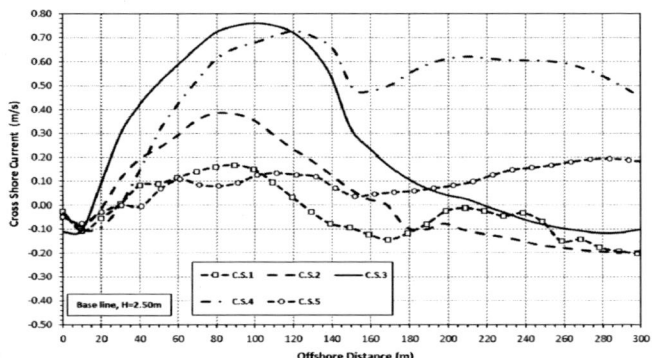

Figure 6.13-e: Cross shore current at the profiles for H=2.50m.

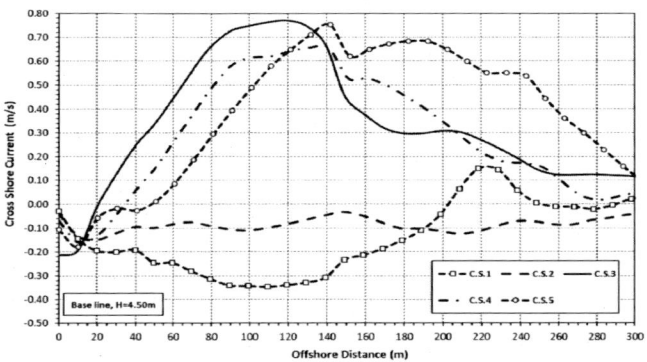

Figure 6.13-f: Cross shore current at the profiles for H=4.50m.

6.4. Impacts of the Perched Beach

Numerical simulations have been made for the case of the actual dimensions of the structures and site specific conditions (Alternative (4), Figure 6.5). It can be observed that two gaps exist in the west and east groins and dredging works have been applied to provide better shape of the beach and nearly uniform bed slope within the perched beach.

Figures 6.14a to 6.14c show the wave direction and height as the waves approach the perched beach from the prevailing wind direction during for H=1.77, 2.5 and 4.5m. It can be observed that the incident wave is partially reflected at the submerged breakwater and small waves are transmitted to the protected area over and through the permeable submerged structures. Over the submerged breakwater, the combined effects of wave shoaling and damping occur due to the change in the water depth and the dynamic interaction with the flow inside the submerged breakwater. The gap allows some wave energy into the perched beach and high waves are formed at its offshore face. Thus, it is recommended to prohibit swimmers from accessing the opening and a strong plastic wire mesh should be employed in the openings to save swimmers from any possible movement through them.

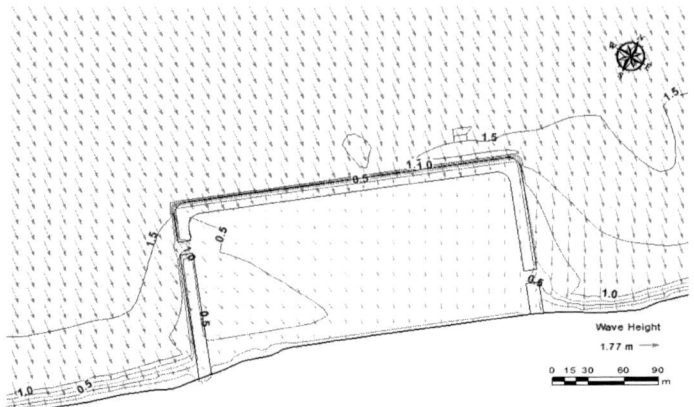

Figure 6.14-a: Computed wave height and direction in the vicinity of the structure for H=1.77m.

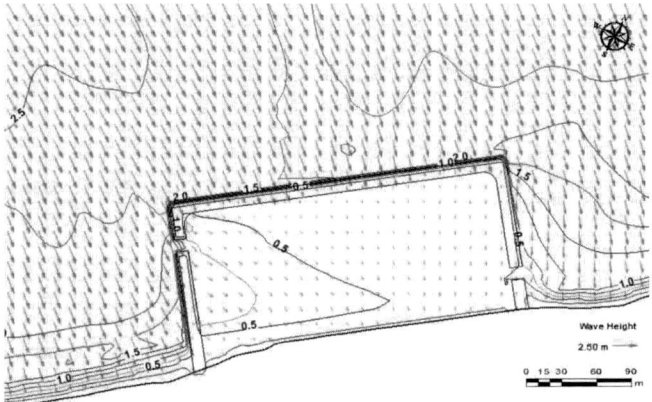

Figure 6.14-b: Computed wave height and direction in the vicinity of the structure for H=2.50m.

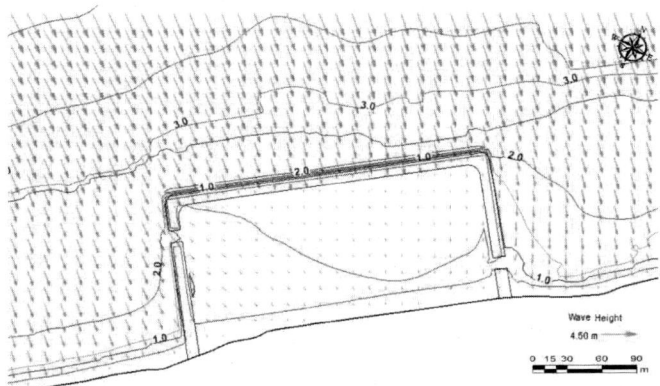

Figure 6.14-c: Computed wave height and direction in the vicinity of the structure for H=4.50m.

It is also evident that wave breaking takes place over the submerged breakwater and small waves are transmitted to the protected area (Figure 6.15). It can be remarked that the transmitted wave height is always less than 0.75m even for the extreme wave height of 4.5m, but it is less than 0.5m for all other incident wave heights conditions. Comparisons between the wave heights for the cases before and after construction of the perched beach are evident in Figure 6.15. It is worthwhile that before the construction of the perched beach, large waves could approach the shoreline at a height of more than 1.0m for the prevailing deepwater summer waves of 1.77m in height. The latter waves break at about 40m from the existing shoreline and the broken wave height is less than 0.5m within the last 20m from the shoreline. This causes inconvenience to swimmers due to the limited width of the surf zone and possible large induced rip currents.

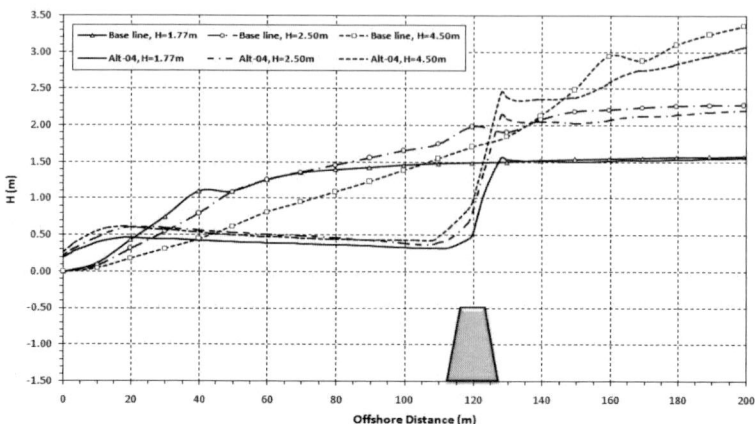

Figure 6.15: Computed wave height at the centerline of the perched beach (at the profile C.S.3).

The virtue of the perched beach is well pronounced as shown in Figure 6.16a to 6.16c as the current velocities in the protected area become very small due to the construction of the perched beach. The velocity inside the perched beach is less than 0.5m/sec which is very convenient for swimmers even during the extreme incident wave conditions. It can be observed that no rip current or circulation occur inside the perched beach, which is safe for swimmers. On the other hand, rip currents and circulation zones are evident outside the perched beach.

The cross and long shore currents along the centerline of the perched beach for various incident wave conditions are presented in Figures 6.17-6.18. It can be noticed that the rip currents is almost null during moderate summer waves, i.e., H=1.77m, and is less 0.2m/s during high summer waves, i.e., H=2.50m. On the other hand, the long shore current inside the perched beach is less than 0.5m/sec during the summer wave. The maximum long shore current velocity reaches 0.70m/s and 0.60m/s during summer for H=2.50m and H=4.50m, respectively, and occur at the offshore toe of the submerged breakwater outside the perched beach. It is also evident that eddies are sometimes generated at the west corner of the submerged breakwater and large current velocities are induced at the offshore toe of the submerged breakwater (Figures 6.16-b and 6.16-c) for the highest summer and winter waves, i.e., H=2.5 and 4.5m. Thus, additional toe protection has been made during construction at the latter locations to save the breakwater from toe scour and possible failure. The currents also cross the opening in the west groin to the perched beach at relatively slow velocity before they reach/ pass the openings in the east groin. Figure 6.16-c shows that two large eddies are generated during the highest winter wave, i.e., H=4.5m, and are located at the west and east sides of the perched beach. Observations have been made after the complete construction of the perched beach and confirmed the aforementioned results of the numerical model to a great extent.

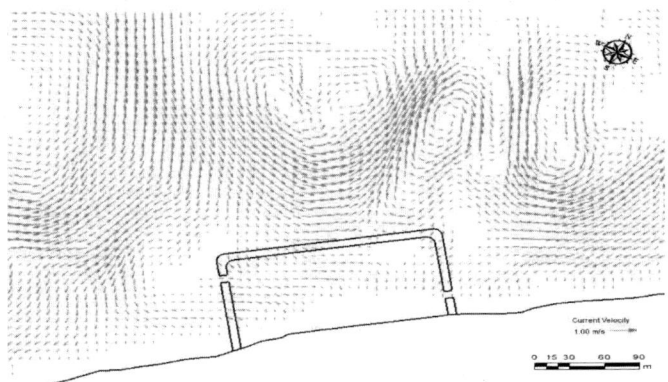

Figure 6.16-a:The current velocity in the vicinity of the structure for H= 1.77m.

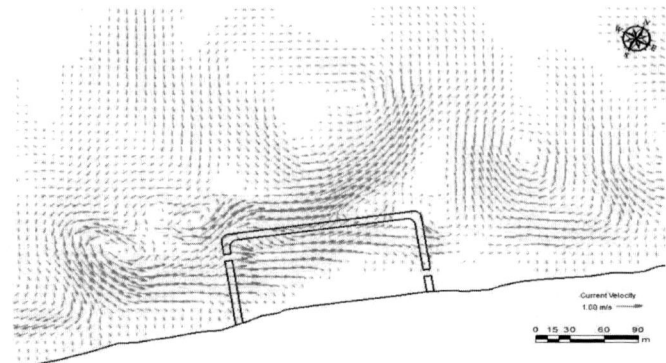

Figure 6.16-b:The current velocity in the vicinity of the structure for H= 2.50m.

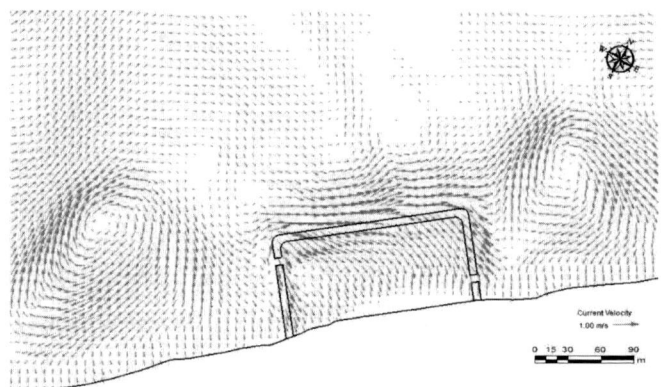

Figure 6.16-c: The current velocity in the vicinity of the structure for H=4.50m.

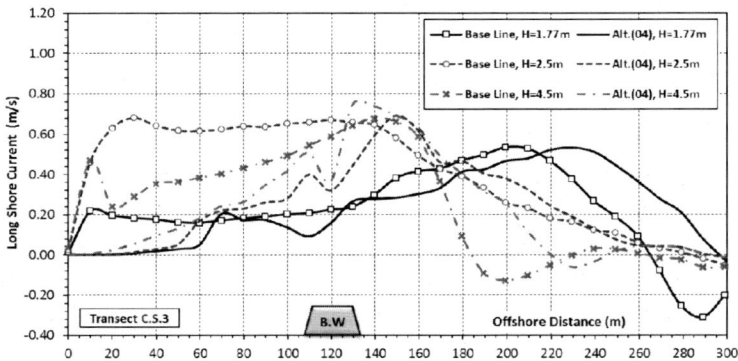

Figure 6.17: Long shore current velocity along C.S.3 before/after construction of the perched beach

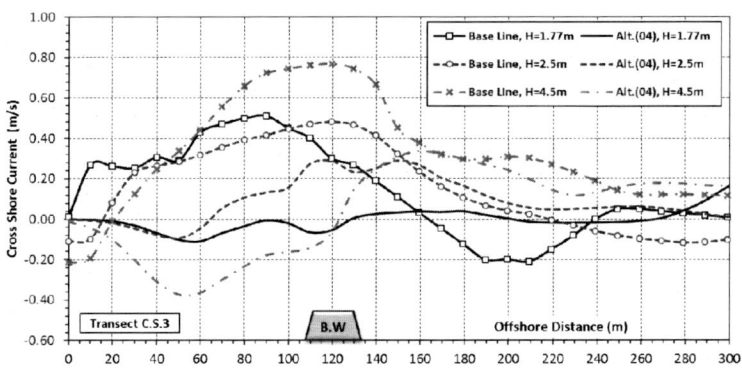

Figure 6.18: Rip current velocity along C.S.3 before/after construction of the perched beach

6.5. Alternatives Discussion

The proposed alternatives have various configurations including: submergence ratio of the breakwater, groin with/without gap and emerged/submerged groin. The alternatives have been compared from the point of view of wave height, currents velocities, flushing and shoreline changes. Different wave conditions were tested, as well, for the different alternatives and the results are presented in appendix A. In the current discussions, it has been focused on the case of maximum deep water wave in summer, i.e., H=2.5m.

6.5.1. Wave Height

Comparisons between the wave heights for the cases before and after construction of the perched beach are evident in Figure 6.19.It is observed that wave breaking takes place over the submerged breakwater and small waves are transmitted to the protected area. These figures confirm that a partial standing wave is formed in front of the breakwater and nonlinear wave damping takes place over it and a smaller wave is transmitted to the onshore side. The results also suggest that wave energy dissipation depends greatly on breakwater height. It is shown that the higher the breakwater is, the lower the transmitted but the slightly higher the reflected wave energies are. It has been found that the transmitted wave height is less than 0.60m for the alternatives: Alt. (01), Alt. (02),Alt. (03), and Alt. (04) which have the same submergence ratio (d/h=0.84), which mean that the site can be used for recreational purposes for most of the summer season. On the other hand, the transmitted wave height is more than 0.6m for the alternatives: Alt. (05), and Alt. (06). Figure 6.21 shows that there is no significant effect on the wave height at the centerline of the perched beach due to the gaps in groin and the emerged/submerged groins.

So, these wave conditions are considered as non-comfortable swimming conditions, as stated by Saski et al (1975)and hence these alternatives do not provide safe swimming conditions.

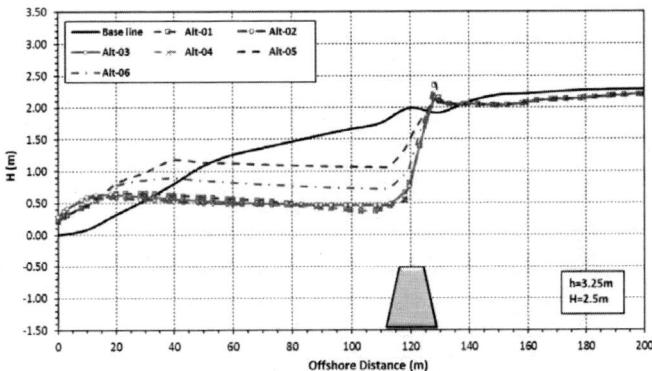

Figure 6.19: Computed wave height along the centerline of the perched beach (H=2.50m).

6.5.2. Current Velocity

The cross and long shore currents along the centerline of the perched beach for the different alternatives are presented in Figures 6.20-6.21during high summer waves, i.e., H=2.50m. It can be noticed that the rip currents before construction of the perched beach is more than 0.7m/s, while, it decreases after construction to 0.2m/s inside the perched beach. Also, it can be noticed that the long shore currents is less than 0.2 m/s for a distance 60m offshore, but, after this distance it increases rapidly to reach 0.46ms/, 0.68m/s and 0.77m/s in case of Alt. (01), Alt. (06) and Alt. (05), respectively, and occur at the offshore toe of the submerged breakwater outside the perched beach. So, it can confirmed that the higher the breakwater is, the lower offshore currents are. Also, attention should be given to design of the breakwater toe. Furthermore, swimming close to the offshore side of the breakwater is forbidden.

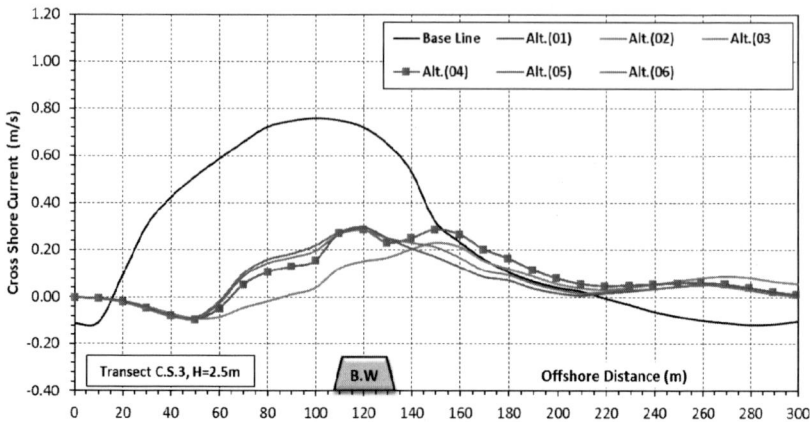

Figure 6-20: Rip current velocity along the perched beach(H=2.50m).

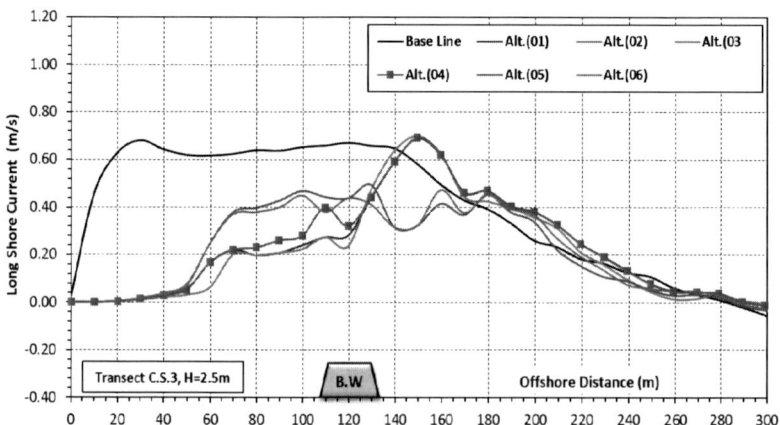

Figure 6-21: Long shore current velocity along the perched beach(H=2.50m).

6.5.3. Shoreline Changes

Computations have been made using GENSIS program within the package of SMS model for shoreline locations. Calibration has been made for the sediment transport model (GENSIS) using the bathymetric survey data conducted in 2009 and 2012 before commencement of construction of the perched beach, and the measured in 2013. Figure 6.22 shows the computed and measured shoreline changes after one year from the perched beach construction. A comparison was prepared between the computed and measured shoreline changes for different cases of different calibration coefficients (K1, K2) in the longshore sediment transport formula. Figure 6.23 illustrate the correlation between the measured shoreline and the computed shoreline for various values of calibration coefficients. The correlation value (R2 =0.7376) for K1=0.65 and K2=0.21. The correlation value (R2 =0.8788) for K1=0.30 and K2=0.05. However, the correlation value (R2 =0.9906) for K1=0.40 and K2=0.08. Thus, the best values of the calibration coefficients for the study area are K1=0.40 and K2=0.08.

According to the proposed the long shore sand transport calibration coefficients, it can be observed that there are small differences between the computed and measured shoreline changes along the up drift and down drift of the structure. Moreover, computations have also been made for predicting the shoreline changes after complete construction of the perched beach being in Jan. 2013 and the results are presented in Figure 6.24 after 3 years later, i.e., 2016. The longshore sediment transport has been calculated based on Ozasa and Brampton formula in SMS model. The net transport has been computed over three years of simulation, including seabed update. The computed net sediment transport rate that corresponds to the period 2009-2012, i.e., before construction of the perched beach, has been found to be 15000 m^3/year and its direction is east ward.

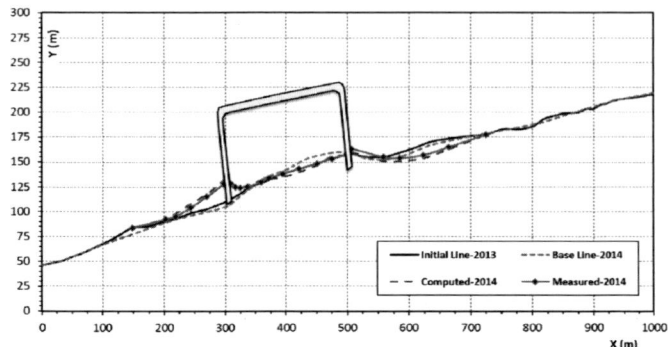

Figure 6-22: Computed and measured shoreline changes around the perched beach.

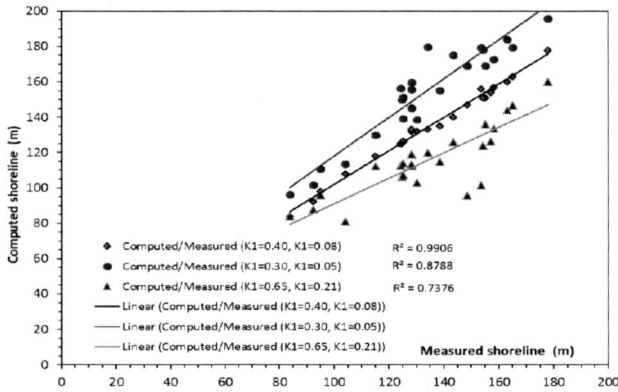

Figure 6-23: Correlation of the computed and measured shoreline changes.

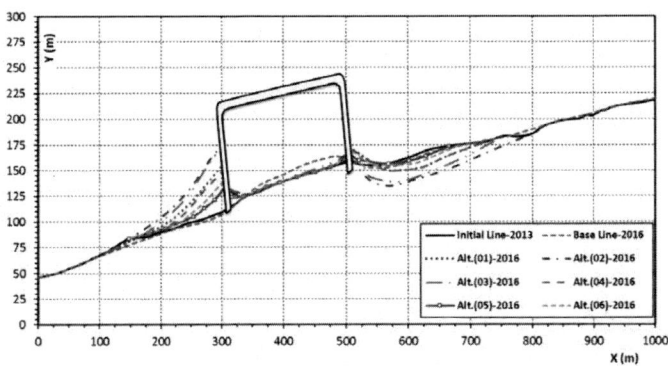

Figure 6.24: Shoreline changes around the perched beach for various alternatives after 3 years of construction as compared with the case without a perched beach.

The shoreline at the up drift side has been anticipated to progress for 13.3m/year seaward in case of no gap (Alt. 01) and for 11.7m/year if the gap is kept clear (Alt. 04) during the period 2013-2016 (Table 6.3). Thus accretion is slightly slower in case of the gap than in case of no gap. It can be stated that the gap allows some sediment to cross the groin into the perched beach and hence less accretion occurs at the up drift side. On the other hand, the down drift side has been found to erode at slow rate. The shoreline at the down drift side has been found to retreat for 4.5m/year in case of no gap (Alt. 01) and for 2.8m/year if the gap is kept clear (Alt.04). As well, erosion is slightly slower in case of the gap than in case of no gap. The accretion rate has been estimated by the numerical model and found to be9600 to 7200 m^3/year at the up drift side. The corresponding erosion rate has been found to be 6850 and 5100 m^3/year for the cases of no gap and gap, respectively. Although the erosion is generally within acceptable limits, the gap seems to considerably reduce the erosion rate and shoreline retreat.

Table 6.3: Accretion and erosion rates around the perched beach for various alternatives.

Alt. - No.	Up drift		Down drift	
	Accretion Rate (m/year)	Accretion Vol. (m3/year)	Erosion Rate (m/year)	Erosion Vol. (m3/year)
Alt. (01)	13.3	9600	4.5	6850
Alt. (02)	20.0	20160	7.2	15900
Alt. (03)	19.3	15120	6.5	11500
Alt. (04)	11.7	7200	2.8	5100
Alt. (05)	6.7	3600	1.3	2050
Alt. (06)	9.3	4800	2.5	2400

Moreover, the submergence ratio has significant effect on the shoreline changes. The shoreline at the up drift side has been found to progress for 13.3m/year, 9.3m/year and 6.7m/year toward the sea in case of d/h=0.84, d/h=0.72 and d/h=0.62, respectively. While, the shoreline at the down drift side has been predicted to retreat for 4.5m/year, 2.5m/year and 1.3m/year toward the shore in case of d/h=0.84, d/h=0.72 and d/h=0.62, respectively. The accretion rate has been found to be 9600m^3/year, 4800m^3/year and 3600m^3/year at the up drift side. The corresponding erosion rate has been found to be 6850m^3/year, 2400m^3/year and 2050m^3/year for the cases of d/h=0.84, d/h=0.72 and d/h=0.62, respectively (Figures 6.25-6.26).

It can be observed that the emerged/submerged groins have significant effect on the shoreline changes along the up drift and down drift of the perched beach, as well. In case of emerged groins (Alt. 02) the shoreline at the up drift side progresses for 20.0m/year and the shoreline at the down drift side retreats for 7.2m/year. In case of submerged groins (Alt. 01), the shoreline at the up drift side progress for 13.3m/year toward the sea and the shoreline at the down drift side retreats for 4.5m/year toward the shore. The accretion rate has been found to be 20160m^3/year and 9600m^3/year at the up drift side for Alt. 02 and Alt.01, respectively. The corresponding erosion rate has been found to be 15900m³/year and 6850m³/year for the cases of emerged groins and submerged groins, respectively. Although the erosion is generally within acceptable limits, the submerged groins/breakwater seems to considerably reduce the erosion rate and shoreline retreat.

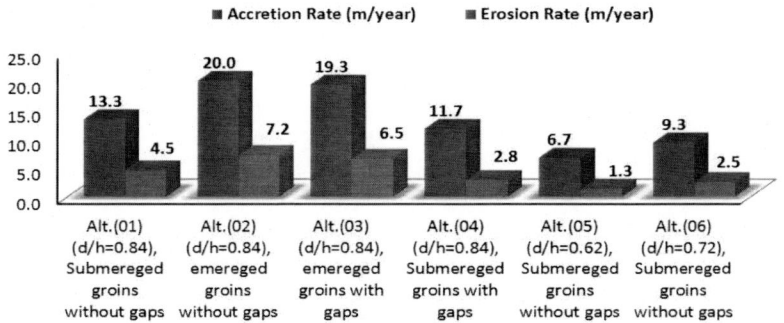

Figure 6.25: Accretion and erosion rates around the perched beach for various alternatives.

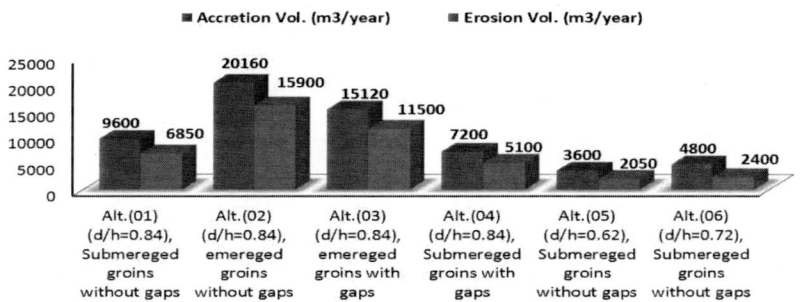

Figure 6.26: Accretion and erosion volume rates around the perched beach for various alternatives.

To minimize and mitigate erosion at the down drift, sand nourishment is considered to be the most straight forward soft approach. Thus, it is possible to move the sand from the accretion zone in the up drift to nourish the erosion area in the down drift. This will allow coarse sand particles deposited at the up-drift to nourish the erosion zone in the down drift instead of the finer particles and hence erosion rate would become less over the time.

For the constructed alternative in the pilot area (Alt. 04), it is recommended to perform the sand transport once every 3 years to limit the maximum shoreline retreat to a maximum of 10m while keeping the gap in good working conditions. It is also evident that the shoreline at the up-drift side is almost unaffected by the perched beach after 100m away from the west groin within the period of 2013-2016, which is less than the groin length. On the other hand, the extent of erosion at the down drift side is limited to about 200m away from the east groin during the period 2013-2016, which is less than twice the groin length. The shoreline in the protected area has been found to be almost unaffected during the period 2013-2016 due to the low currents. It is noteworthy that the spacing between the groins in the protected area in 2013 and 2016 has been artificially made by dredging early in 2013 to adjust the beach profile in the protected area. The annual nourishment to mitigate the erosion at the down drift, is estimated to be 6300 m^3. The grain size and schedules for the nourishment should be carefully prepared taking into account the social impact.

6.5.4. Flushing Rates

Water quality within lagoons or closed basins must be considered, and it is particularly important for health and environmental quality, especially in warmer climates where biological processes are accelerated. Successful control of water quality is usually dependent upon periodic exchange of the basin water with the sea water of the open sea. Thus, further computations have been made for the flushing rate using RMA models based on the relation between the tidal range and the structure height. According to Goshow et al. (2008), the suitable flushing time for the proposed perched beach area is 5 days.

Figure 6-27 shows the time series for concentration of tracer material for various alternatives at station (B). The flushing times are about 7.5 days, 11 days and 9.5 days in cases of alternatives: Alt. (01), Alt. (02) and Alt. (03), respectively. These values exceed the target value of 5 days. However, the flushing times are about 5 days, 2.5 days, and 3 days in cases of alternatives: Alt. (04),Alt. (05) and Alt. (06), respectively. These values are less than the target value of 5 days.

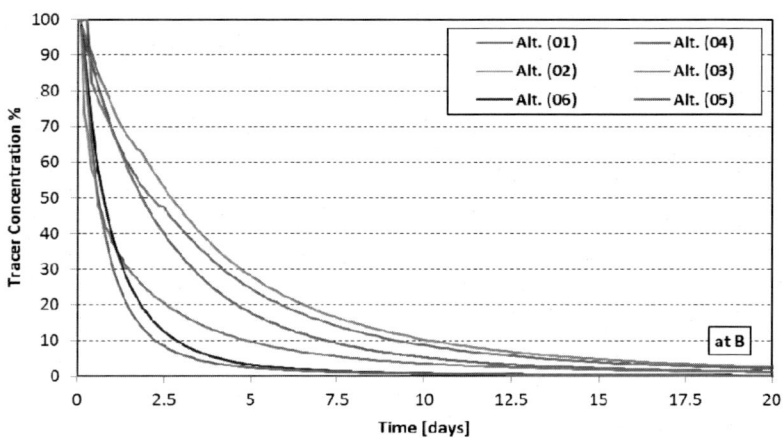

Figure 6.27:Time series for concentration of tracer material for various alternatives.

6.6. Assessment of Alternatives

The alternatives have been simulated, the results analyzed and compared from the point of view of wave height, currents velocities, flushing and shoreline changes, to provide general guidelines for the design of a perched beach within the Al-Arab Bay zone in Cell 5 of the North West coast of Egypt. Table 6.4 shows the comparison among alternatives.

Saski et al (1975) conclude that the comfortable swimming conditions are formed for breaker heights smaller than 0.6m and current velocities smaller than 0.2m/s. Furthermore, the required flushing time is 5 days. Based on the stated comfortable swimming conditions and the data in Table 6.4, it can be observed that the transmitted wave height and the currents velocities are acceptable for the alternatives: Alt. (01), Alt. (02),Alt. (03), and Alt. (04). On the other hand, the transmitted wave height and the currents velocities are unacceptable for the alternatives: Alt. (05) and Alt. (06).

Although, the shoreline changes are minimum in case of alternatives: Alt. (05) and Alt. (06), however, these alternatives are unacceptable for the transmitted wave height and current velocities. Moreover, the shoreline changes in case of alternatives: Alt. (05) and Alt. (06) are larger than the other alternatives.

Also, the flushing times are unacceptable in cases of alternatives: Alt. (01), Alt. (02) and Alt. (03). However, the flushing times are acceptable in cases of alternatives: Alt. (04),Alt. (05) and Alt. (06).

Table 6.4: Comparison of alternatives.

Alt.- No.	Wave height (m)	Currents velocities (m/s)		Shoreline changes (m/year) Accretion/Erosion	Flushing rate (days)
		Rip	Long shore		
Base Line	>1.0	>0.7	>0.68	Stable	N.A
Alt. (01)	<0.6	<0.2	<0.2 for a distance 60m offshore, while 0.2-0.44 until the breakwater	13.3/4.5	7.5
Alt. (02)	<0.6	<0.2	<0.2	20.0/7.2	11
Alt. (03)	<0.6	<0.2	<0.2	19.3/6.5	9.5
Alt. (04)	<0.6	<0.2	<0.2 for a distance 60m offshore, while 0.2-0.46 until the breakwater	11.7/2.8	5
Alt. (05)	>0.6	<0.2	0-0.77	6.7/1.3	2.5
Alt. (06)	>0.6	<0.2	0-0.68	9.3/2.5	3
Comfortable swimming	<0.6	<0.2	<0.2	Minimum	5

conditions (Saski et al, 1975)					(Goshow et al., 2008)

It can be judged that the configurations of alternatives: Alt. (01) and Alt. (04) are the most suitable, but still Alt. (04) is the better due to the groin gaps effectiveness. Thus, it can be considered that the configurations of alternative (04) provide guidelines for the design of a perched beach within the Al-Arab Bay zone in Cell 5 of the North West coast of Egypt.

6.7. Conclusions

A study was carried out to evaluate a pilot perched beach project recently constructed along the coast of the Arab bay in Egypt. Six alternatives of perched beaches as necessary to develop safe swimming conditions have been studied. Investigations have been made for the case before and after construction of the perched beach to evaluate the impacts of the pilot project using SMS numerical model. The followings have been concluded:

- Numerical results and initial field observations have shown that the constructed perched beach could be a reliable solution for protecting swimmers along the North-West coast of Alexandria in Egypt. The impacts on flushing and water quality seem to be acceptable especially with the aid of openings in the groins. The impacts of groins/breakwater on shoreline changes have been predicted and they are tolerable at reasonable annual costs.

- The configurations of alternatives: Alt. (01) and Alt. (04) are the most suitable, but still Alt. (04) is the best due to the groin gaps effectiveness. So, it can be considered that the configurations of alternative (04) provide guidelines for the design of a perched beach along the coast of the Arab bay in Egypt.

- For the constructed perched peach (Alt.04), it has been found that:
    - The maximum length of erosion in the down drift (east of the jetties) is less than twice the jetty length and the shoreline retreats at less

than 3m/year, but the maximum accretion length is less than the jetty length and the shoreline progresses at less than 12m/year. The rates of sediment transport up drift and down drift the perched beach are 7200 and 5100 m^3/year, respectively, and sand nourishment is recommended once every three years. The residual adverse impacts could be generally alleviated through routine sand nourishment program of the shoreline.

- The flushing times is about 5 days. This value is in a good agreement with the required flushing time according to Goshow et al., 2008.

Chapter 7
Conclusions and Recommendations

7.1. General

Coastal zones in many countries suffer from strong offshore direct rip currents. Rip currents regularly lead to hazardous situations, and at some beaches swimming is prohibited for a considerable time of the year especially during summer storm. Also, the swimming may be prohibited at beaches where the wave height is too high. Saski et al (1975) concluded that breaker heights smaller than 0.6m and current velocities smaller than 0.2m/s are considered as comfortable swimming conditions, but it is hard to swim against a rip current of 0.5m/s and breaker height greater than 2.0m even for good swimmers. So, there is an essential need to construct suitable coastal structure in order to secure safe conditions for swimmers.

The present study has been recommended the use of perched beach as a possible alternative for safe swimming conditions. Numerical simulation has been implemented by using the Surface Water Modeling System (SMS-10.1). The numerical model was applied to investigate various configurations of the perched beach including submergence ratio of the breakwater, groin with/without gap, the gap width/location and emerged/submerged groin. These configurations have been compared from the point of view of wave height, currents velocities, flushing rates and shoreline changesto develop general guidelines for the design of similar constructions.

An actual scale model of a perched beach designed and constructed to provide a safe swimming conditions. The project area is constructed along the North-West coast of Alexandria in Egypt. The latter area has long been suffered from rip currents as large as 0.7m/sec and limited safe swimming strip of less than 40m during the prevailing wave conditions in the summer season. Six alternatives of perched beaches have been studied using SMS numerical model and adopting the actual wave rose of El-Dekhila port and bathymetric survey of the project area. Field measurements have been carried out before construction of the perched beach, and it is considered as the baseline condition. The model has been calibrated and validated against the collected and measured field data. The alternatives have been simulated, the results analyzed and compared to evaluate the pilot perched beach project. Special attention has been given to alleviate the possible adverse impact of the protection structures. These impacts include excessive shoreline erosion and possible deterioration of water quality.

7.2. Conclusions

The main conclusions are noted as follows:

- Three numerical models, i.e., coupled CMS-Wave and BOUSS-2D model, GENESIS model and RMA models, have been found to reproduce satisfactorily the hydrodynamic conditions within perched beaches.

- The perched beach could be a reliable solution for protecting swimmers along the coast with minimum impact on the shoreline while preserving acceptable water quality within the perched beach. Good flushing rates can be achieved if the crest level of the submerged breakwater and groins is at least 0.5m below M.S.L. and the groin at the up drift side has a gap located near its offshore end, while the groin at the down drift side has a gap close to the shoreline. To allow water flow along the shoreline, it is recommended to have the up drift gap wider than the down drift one. Attention should be given that no attendance is permitted near the gaps due to possible high current velocities and eddies.

- Application has been made to the North-West coast of Egypt and it has been found that the appropriate configurations of the perched beach to provide acceptable flushing and minimum impact on the shoreline is as follows:

  o The crest level of the submerged breakwater and groins should be set at 0.5m below M.S.L.

o The gap width in the up drift groin shall be 10% of the groin length and constructed offshore at 75% of the groin length measured from the shoreline.

o The gap width in the down drift groin shall be 20% of the groin length and constructed at 25% of the groin length measured from the shoreline.

- Large current velocities and eddies are generated offshore the submerged breakwater and offshore the groin at the up drift side. The need for extra toe protection at these locations is far evident than other locations. Attention should be given to bed scour and toe design.

- Considerable shoreline changes have occurred along the shoreline in the vicinity of the perched beach. These changes diminish within three times the length of the groin on both sides of the perched beach.

- Flushing process is slightly higher in case of the gaps than in case of no gap. However, attention should be made that higher flushing is caused by higher waves and currents in the perched beach and this may affect the convenience of swimmers and may lead to possible dragging of swimmers out of the perched beach. A compromise should though between the need for flushing and convenience/safety of swimmers in the perched beach.

7.3. Recommendations

Based on the current study, it is recommended to:

- Study the effect of the perched beach area and groins shape/angle on wave height, current velocities, flushing rates and shoreline changes.

- Study the effect of the internal properties of breakwater/groins on wave height, current velocities, flushing rates and shoreline changes.

- Study and develop mitigation measures of trapping floats and sea weeds/grasses inside the perched beach.

REFERENCES

1. Artagan, S.S., (2006). "A One Line Numerical Model for Shoreline Evolution Under the Interaction of Wind Waves and Offshore Breakwaters", M.S. Thesis, Ankara.

2. Bailard, J. A. (1984). "A Simplified Model for Longshore Sediment Transport" 19th International Conference on Coastal Engineering, ASCE, Texas, 1454-1470.

3. Bakker, W.T., (1968). "The Dynamics of a Coast with a Groin System", Proc. 11th Int. Conf. on Coastal Engrg., ASCE, London, pg. 492-517.

4. BCEOM and L.C.H.F. El Dikheila Port Project Studies, June 1977

5. Birkemeier, W. A. (1985). "Field Data on Seaward Limit of Profile Change", Journal of Waterway, Port, Coastal and Ocean Engineering 111 (3), 598-602.

6. Booij, N., R. C. Ris, and L. H. Holthuijsen, (1999). "A Third-Generation Wave Model for Coastal Regions", Journal of Geophysical Research 104 (C4): 7,649-7,666.

7. Capobianco, M., Hanson, H., Larson, M., Steetzel, H., Stive, M.J.F., Chatelus, Y., Aarninkhof, S., Karambas, T., (2002). "Nourishment Design and Evaluation: Applicability of Model Concepts", Coastal Engineering, Vol. 47, pg. 113-135.

8. Choi, K.W., Lee, J.H.W., and Cheung, Y.K., (1989). "A Numerical Study of Tidal Flushing in a Typhoon Shelter", Proc., 4th Asian Fluid Mech. Congress, Vol. 1, (edited by N.W.M. Ko and S.C.Kot), B32-B35.

9. d'Angremond, K., van der Meer, J., and De Jong, R. (1996). "Wave Transmission at Low-Crested Structures", Proc. 25th Coastal Engineering Conference, ASCE, pp. 3305-3318.

10. Dabees, M. A., and Kamphus, J. W. (2000). "NLINE; Efficient Modeling of 3-D Beach Change", 25th International Conference on

Coastal Engineering, Sydeny-Australia, 2700-2713.

11. Dally, W. R., and Dean, R. G. (1985). "Wave Height Variation Across Beaches of Arbitrary Profile", Journal of Geophysical Research, 90 (C6), 917-927.

12. Demirbilek, Z., A. Zundel, and O. Nwogu. (2005a). "BOUSS-2D Wave Model in SMS: I. Graphical Interface", Tech. Note ERDC/CHL-I-69, U.S. Army Engineer R&D Center Vicksburg, MS.

13. Demirbilek, Z.; Lin, L., and Seabergh, W.C., 2009. "Laboratory and Numerical Studies of Hydrodynamics Near Jetties", Coastal Engineering Journal 51 (2): 143-175 JSCE.

14. Demirbilek, Z. and J. D. Rosati. 2011. "Verification and Validation of the Coastal Modeling System", Report I, Executive Summary. Tech. Report ERDC/CHL-TR-11-xx, U.S. Army Engineer Research and Development Center, Coastal and Hydraulics Laboratory, Vicksburg, MS.

15. E.C.S.I., Coastal currents measurements for area kilometers 50.75 to 51.50 along Alexandria/ Mersa Matruh, 1980.

16. Fischer, H.B., List, E.J., Koh, R.C., Imberger, J., Brooks, N., "Mixing in Inland and Coastal Waters", Academic Press, 1979.

17. Fleming, C.A. and Hunt, J.N., (1976), "Application of a Sediment Transport Model", Proc. 15th Int. Conf. on Coastal Engrg., ASCE, pg.1184-1202.

18. Frihy, O. E. (1992). "Sea-Level Rise and Shoreline Retreat of the Nile Delta Promontories, Egypt", Natural Hazards, 65-81.

19. Frihy, O.E, Nasr, S.M, El Hattab, M.M., M. E. R. (1994). "Remote Sensing of Beach Erosion Along the Rosetta Promontory, Northwestern Nile Delta, Egypt", International Journal of Remote Sensing, 15, 1649-1660.

20. Frihy, O. E, Deabes, E. A., Gindy, A. A. E., and Shallalat, E. (2010). "Wave Climate and Nearshore Processes on the Mediterranean Coast of Egypt", 103-112.

21. Goda, Y. and Ahrens, J.P. (2008). "New Formulation for Wave

Transmission Over and Through Low Crested Structures", Proc. 31st
Coastal Engineering Conference, ASCE, pp. 3530–3541.

22. Goshow, C.K., S.E. Hosseini, & S. O'Neil (2008). Waterfront Developments in the Gulf Region: A review and examination of flushing and its relevance to water quality. Seventh International Conference on Coastal and Port Engineering in Developing Countries, (COPEDEC VII), Dubai, 24-28 Feb. Paper No. 59.

23. Hallermeier, R. J., and Belvoir, F. (1978). "Uses for a Calculated Limit Depth to Beach Erosion", Coastal Engineering, 1493-1512.

24. Hallermeier, R. J. (1981). "A Profile Zonation for Seasonal Sand Beaches from Wave Climate", Coastal Engineering, 4, 253-277.

25. Hanson, H., Kraus, N.C., (1986a). "Seawall Boundary Condition in Numerical Models of Shoreline Evolution", Technical Report CERC-86-3, U.S. Army Engineer Waterways Experiment Station, Vicksburg, MS.

26. Hanson, H., and Kraus, N.C., (1986b). "Seawall Constraint in Shoreline Numerical Model", Journal of Waterway, Port, Coastal and Ocean Engineering, Vol.111, No.6, pg.1079-1083.

27. Hanson, H., (1987). "GENESIS: A Generalized Shoreline Change Numerical Model for Engineering Use", Ph.D. Thesis, University of Lund, Lund, Sweden.

28. Hanson, Hans, and Kraus, N. C. (2011). "Long-Term Evolution of a Long-Term Evolution Model", Journal of Coastal Research, 59 (1989), 118-129.

29. Horikawa, K., and Kuo, C.-T. (1966). "A Study on Wave Transformation Inside Surf Zone", Coastal Engineering, New York, 217-233.

30. John B. Herbich, (1996). "Coastal and Ocean Engineering", Volume I: Wave Phenomena and Coastal Structures, pages 591-634

31. Kamphuis, J W, Readshaw, J. S., Canada, W., and Laboratories, H. (1978). "A Model Study of Along Shore Sediment Transport Rate", 21st International Conference on Coastal Engineering, ASCE, Malaga,

1253-1264.

32. Kamphuis, J W, Davies, M. H., Nairn, R. B., and Sayao, O. J. (1986). "Calculations of Littoral Sand Transport Rate", Coastal Engineering, 10, 1-21.

33. King DB (2005). "Influence of grain size on sediment transport rates with emphasis on the total long-shore rate", US Army Corps, 2005.

34. King, I.P. (1982). "A Three Dimensional Model for Stratified Flow", Proceedings of the 4th International Symposium on Finite Elements in Flow, Tokyo, Japan.

35. King, I.P. (1985). "Strategies for Finite Element Modelling of Three Dimensional Hydrodynamic Systems", Advanced Water Resources , Vol. 8, June, pp. 69-76.

36. King, I.P. and J.F. De George, (1995). "A Multi-Dimensional Modeling of Water Quality Using the Finite Element Method", 4th International Conference on Estuarine and Coastal Modeling, ASCE, October.

37. LeMéhauté, B., and Soldate, M., (1978). "Mathematical Modeling of Shoreline Evolution", Proc. 16th Int. Conf. on Coastal Engrg., ASCE, pg.1163-1179.

38. Lin, L., Z. Demirbilek, and F. Yamada, (2008),"CMS-Wave: A Nearshore Spectral Wave Processes Model for Coastal Inlets and Navigation Projects", Coastal and Hydraulics Laboratory Technical Report ERDC/CHL TR-08-13.

39. Lin, L., Z. Demirbilek, and H. Mase, (2011). "Recent Capabilities of CMS-Wave: A Coastal Wave Model for Inlets and Navigation Projects", Proceedings, Symposium to Honor Dr. Nicholas Kraus. Journal of Coastal Research, Special Issue 59, 7-14.

40. Luff, R. and Pohlmann, T., (1995). "Calculation of Water Exchange Times in the ICES- Boxes with a Eulerian Dispersion Model Using a Half-Life Time Approach", Deutsche Hydr. Zeitschr, Vol. 47, 287-299.

41. Madsen, P. a., Fuhrman, D. R., and Wang, B. (2006). "A Boussinesq-

Type Method for Fully Nonlinear Waves Interacting with a Rapidly Varying Bathymetry", Coastal Engineering, 53 (5-6), 487-504.

42. Mase, H. (2001). "Multi-Directional Random Wave Transformation Model Based on Energy Balance Equation", Coastal Engineering Journal, 43 (04), 317-337.

43. Mase, H., Oki, K., Hedges, T. S., and Li, H. J. (2005). "Extended Energy-Balance-Equation Wave Model for Multidirectional Random Wave Transformation", Ocean Engineering, 32(8-9), 961-985.

44. Monsen, N.E., Cloern, J.E. and Lucas, L.V., (2002). "A Comment on the Use of Flushing Time, Residence Time and Age as Transport Time Scales", Limnology and Oceanography, Vol.47, No. 5, 1545-1553.

45. Nafaa, M.E. and Frihy, O. E. (1993). "Beach and Near Shore Features Along the Dissipative Coastline of the Nile Delta, Egypt", J. Coastal Res., 9, pp. 423-433.

46. Nwogu, O., (1993). "An Alternative Form of the Boussinesq Equations for Modeling the Propation of Waves Form Deep to Shallow Water", Journal of Waterway, Port, Coastal and Ocean Engineering, ASCE, 119 (6), pp. 618-638.

47. Nwogu, O. (1993). "Alternative Form of Boussinesq Equations for Nearshore Wave Propagation", J. Wtrwy., Port, Coast, and Oc. Engrg., 119: 618-638.

48. Nwogu, O. (1996). "Numerical Prediction of Breaking Waves and Currents with a Boussinesq Model", Proc. 25th Int. Conf. on Coast. Engrg, Vol. 4, ASCE, 4807-4820.

49. Nwogu, O., and Z. Demirbilek. (2001). "BOUSS-2D: A Boussinesq Wave Model for Coastal Regions and Harbors", Coastal and Hydraulics Laboratory Technical Report ERDC/CHL TR-01-25.

50. Oliveira, A. and Baptista, A.M., (1996). "Diagnostic analysis of estuarine residence times", Computational Method in Surface Flow and Transport Problems, (edited by A.A. Aldama, J. Aparicio, C.A. Brebbia, W.G. Gray, I. Herrera and G.F. Pinder), Computational

Mechanics Publication, Southampton, 355-362.

51. Omar, W. M., El-mooty, M. M. A., and Nagy, H. M. (2005). "Evaluation of Closure Depth Along the Northern Nile Delta Coast, Egypt", Alexandria Engineering Journal, 44, 623-635.

52. Ozasa, H, and Brampton, A.H. (1980). "Mathematical Modelling of Beaches Backed by Seawalls", Coastal Engineering, 4, 47-63.

53. Pelnard-Considere, R. (1956). "Essai de Theorie de l'Evolution des Formes de Rivage en Plages de Sable et de Galets", 4th Joumees de l'Hydraulique, Les Energies de la Mar, Question HI, Repport No.1, 289-298.

54. Peregrine, D. H. (1966). "Long Waves on a Beach", Journal of Fluid Mechanics, 27 (4), 815-827.

55. Prandle, D., (1984). "A Modelling Study of the Mixing of ^{137}Cs in the Seas of the European Continental Shelf", Phil. Trans. R. Soc. Lond. A, Vol. 310, 407-436.

56. Şafak, I., (2006). "Numerical Modeling of Wind Wave Induced Longshore Sediment Transport", M.S. Thesis, METU, Ankara.

57. Saski, Tamio, Horikawa, Kiyoshi (1975). "Nearshore Current System on a Gently Sloping Bottom", Coastal Eng. in Japan, Elsevier, 18, pp. 123-142.

58. Sharaf El-Din, S.H. (1974). "Longshore Sand Transport in the Surf Zone Along the Mediterranean Egyptian Coast Limnology and Oceanography", Journal of the Egyptian Society of Engineers, Cairo University, 19, No. 2.

60. Shore Protection Authority (2002). "Integrated Development of Egypt's Northwestern Coastal Zone, Development of Near Shore Water Conditions" A Report prepared by Delft Hydraulics to the Ministry of Water Resources and Irrigation of Egypt.

61. Signell, R.P. and Butman, B., (1992). "Modelling Tidal Exchange and Dispersion in Boston Harbour", Journal of Geophysical Research-Oceans, Vol. 97 (C10), 15591-15606.

62. Smith, J. M. (2001). "Modeling Nearshore Transformation with

STWAVE", Coastal and Hydraulics Laboratory Special Report ERDC/CHL SR-01-01. Vicksburg, MS: U.S. Army Engineer Research and Development Center.

63. Stommel, H. and Farmer, H.G., (1952). "On the Nature of Estuarine Circulation", Part II, Woods Hole Oceanographic Institution, Tech. Rep.WHOI-52-51, Woods Hole, Massachusetts, USA, 131 pp.

64. Swart, D. H. (1976). "Predictive Equations Regarding Coastal Transports", 15th International Conference on Coastal Engineering, ASCE, Hawaii, 1113-1132.

65. Tanaka, N. (1976). "Wave Deformation and Beach Stabilization Capacity of Wide Crested Submerged Breakwaters", Proc. 23rd Japanese Conf. Coastal Eng., JSCE, pp. 152-157.

66. Tajima, Y., and Madsen, O. S. (2006). "Modeling Near-Shore Waves, Surface Rollers, and Undertow Velocity Profiles", Journal of Waterway, Port, Coastal, and Ocean Engineering, (December), 429-438.

67. Tauman J., Enclosing Scheme for Bathing-Beach Development, Coastal Engineering Conference, 1976.

68. Van der Meer, J.W. and Daemen, I.F.R. (1994). "Stability and Wave Transmission at Low-Crested Rubble-Mound Structures", J. W. Way, Port, Coastal and Ocean Eng., ASCE, 120. pp.l-19.

69. Van der Meer, J. W., Briganti, R., Zanuttigh, B., and Wang, B. (2005). "Wave Transmission and Reflection at Low-Crested Structures: Design Formulae, Oblique Wave Attack and Spectral Change", Coastal Engineering, 52, pp. 915-929.

70. Wei, G. and Kirby, J.T., (1995). "Time-Dependent Numerical Code for Extend Boussinesq Equations", Journal of Waterway, Port, Coastal and Ocean Engineering, ASCE 121 (6), pp. 251-263.

71. WL|Delft Hydraulics (2002). "Integrated Development of Egypt's Northwestern Coastal Zone: Design for Pilot Area I", Interim Report 2, A Report prepared by Delft Hydraulics to the Ministry of Water Resources and Irrigation of Egypt.

72. WL|Delft Hydraulics, (2003). "Integrated Development of Egypt's Northwestern Coastal Zone, Development of Near Shore Water Conditions", Report H3791, A Report prepared by Delft Hydraulics to Hydraulics Research Institute (HRI).

Appendix A
Impact of the Perched Beach

Results of Alternative (01)

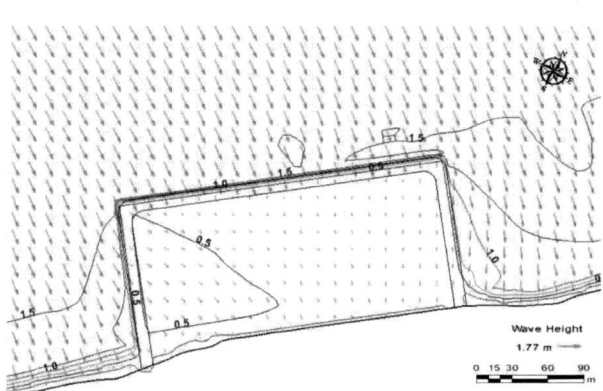

Figure A-1-a: Computed wave height and direction (H=1.77)m.

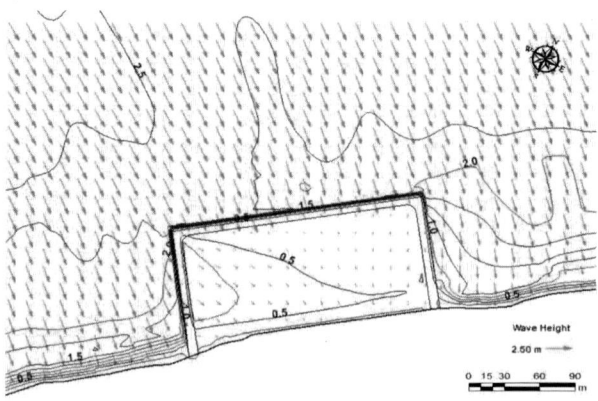

Figure A-1-b: Computed wave height and direction (H=2.50m).

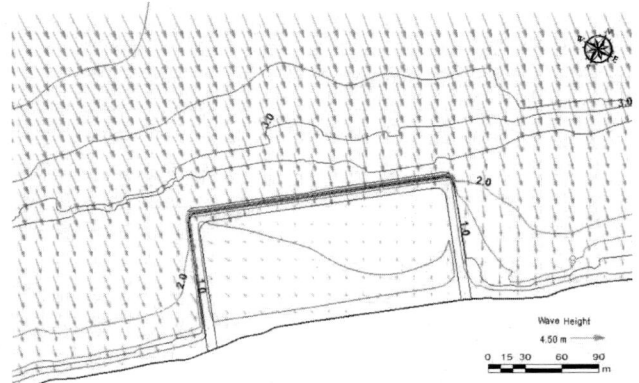

Figure A-1-c: Computed wave height and direction (H=4.50m).

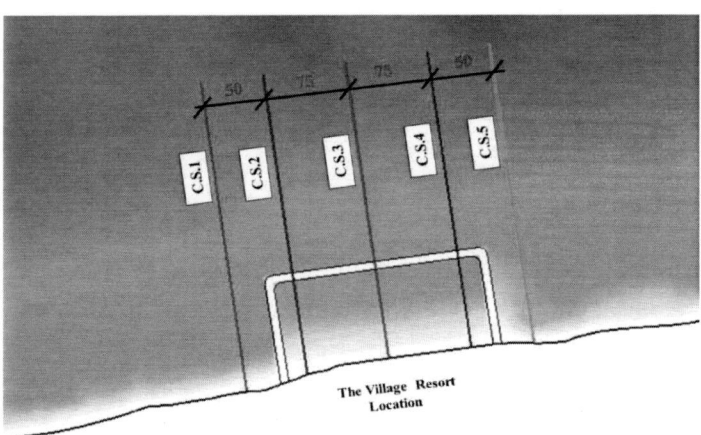

Figure A-2: Location of the transects with respect to the perched beach.

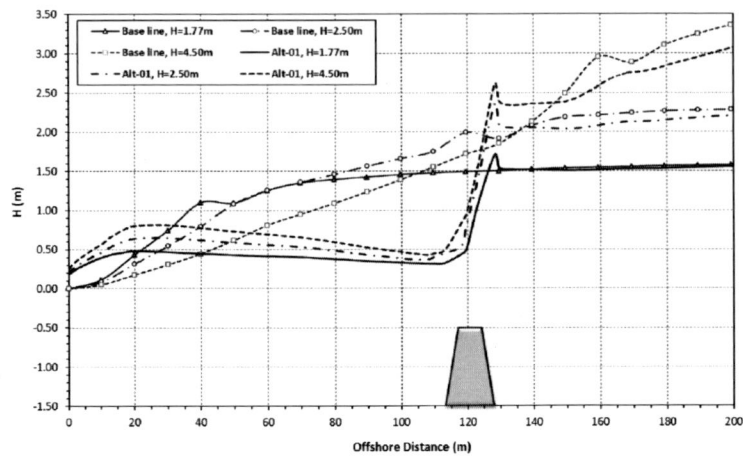

Figure A-3: Computed wave height along the centerline of the perched beach (at C.S.3).

Figure A-4-a: Current velocity in the vicinity of the perched beach (H=1.77m).

Figure A-4-a:Current velocity in the vicinity of the perched beach (H=2.50m).

Figure A-4-a:Current velocity in the vicinity of the perched beach (H=4.50m).

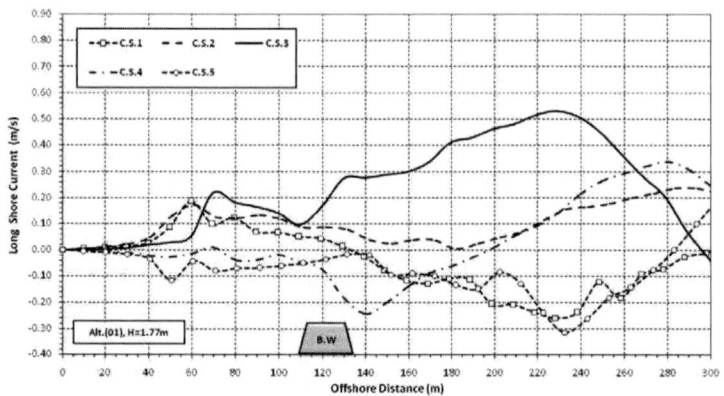

Figure A-5-a:Long shore current velocity along the transects (H=1.77m).

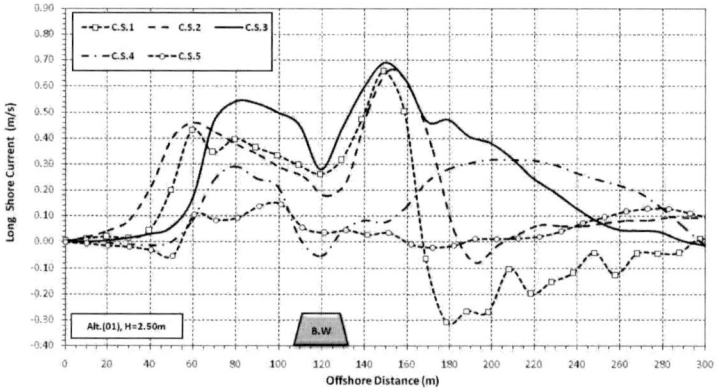

Figure A-5-b:Long shore current velocity along the transects (H=2.50m).

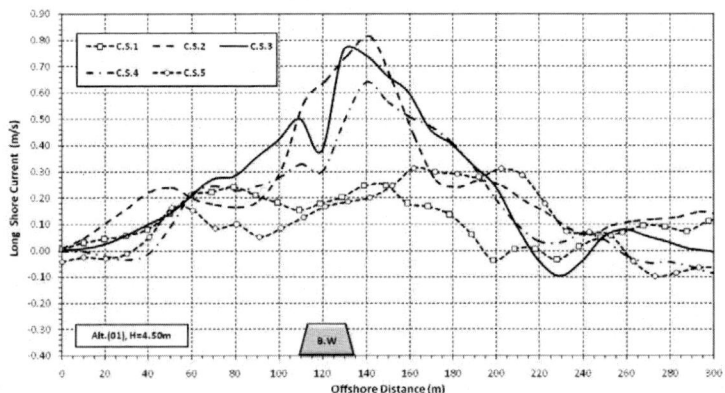

Figure A-5-c:Long shore current velocity along the transects (H=4.50m).

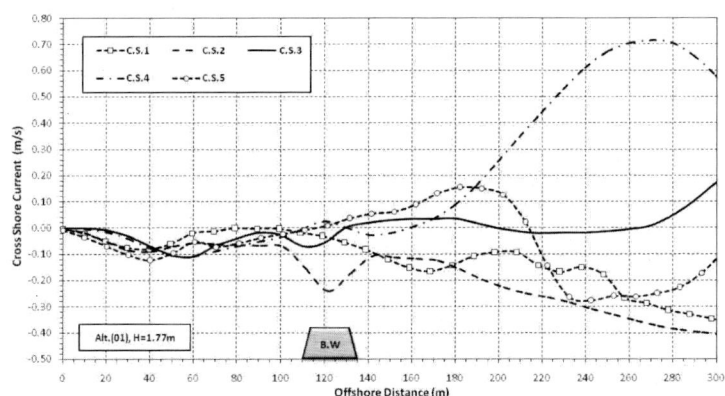

Figure A-6-a:Cross shore (rip) current velocity along the transects (H=1.77m).

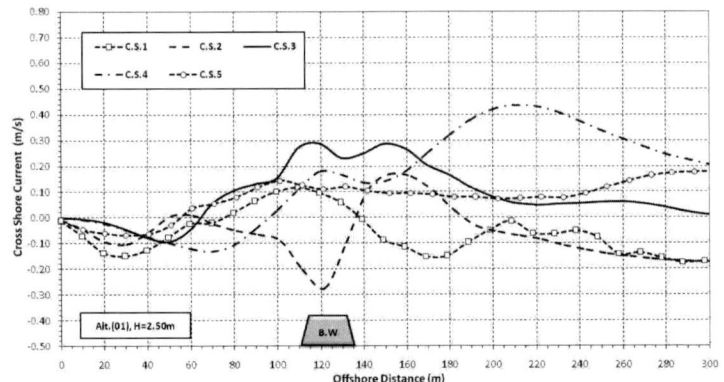

Figure A-6-b:Cross shore (rip) current velocity along the transects (H=2.50m).

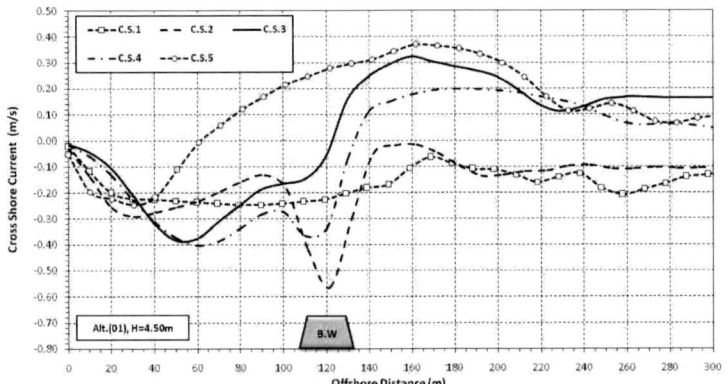

Figure A-6-c: Cross shore (rip) current velocity along the transects (H=4.50m).

Results of Alternative (02)

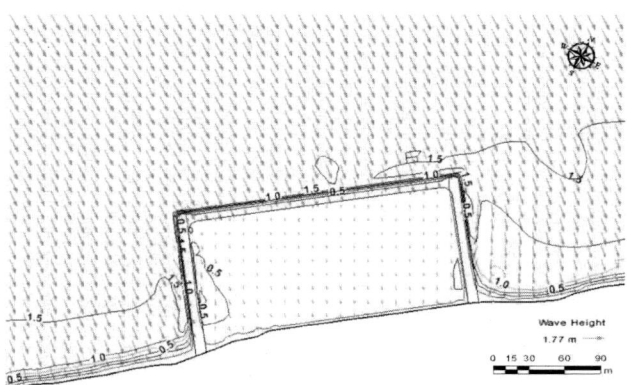

Figure A-7-a: Computed wave height and direction (H=1.77m).

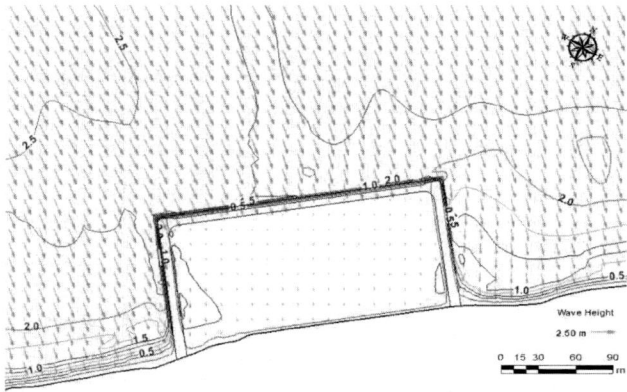

Figure A-7-b: Computed wave height and direction (H=2.50m).

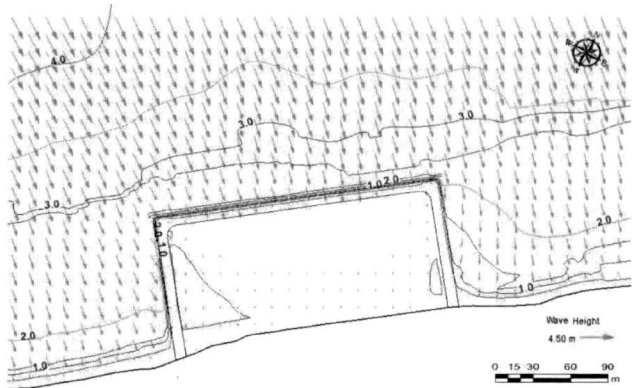

Figure A-7-c: Computed wave height and direction(H=4.50m).

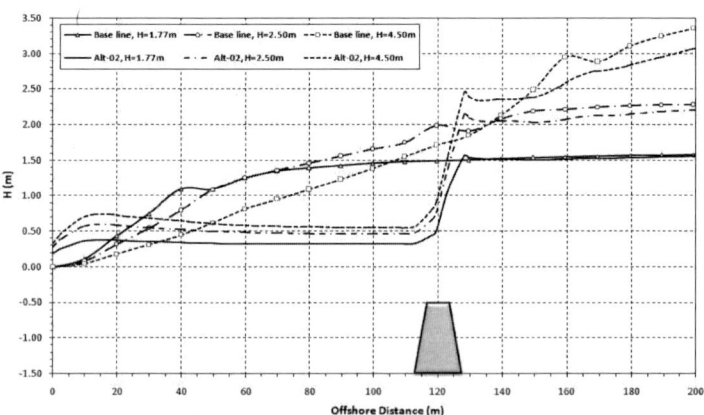

Figure A-8:Computed wave height along the centerline of the perched beach (at C.S.3).

Figure A-9-a:Current velocity in the vicinity of the perched beach (H=1.77m).

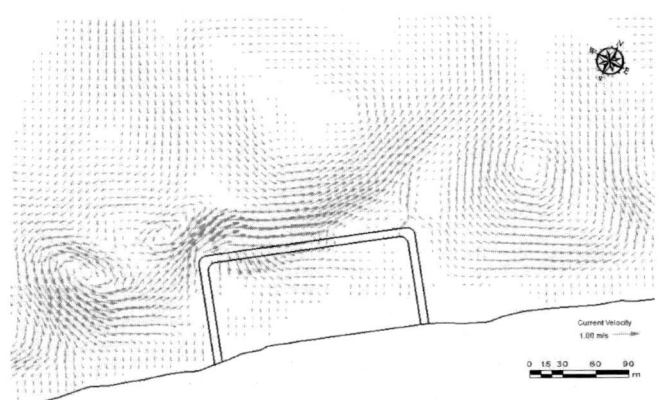

Figure A-9-b:Current velocity in the vicinity of the perched beach (H=2.50m).

Figure A-9-c: Current velocity in the vicinity of the perched beach (H=4.50m).

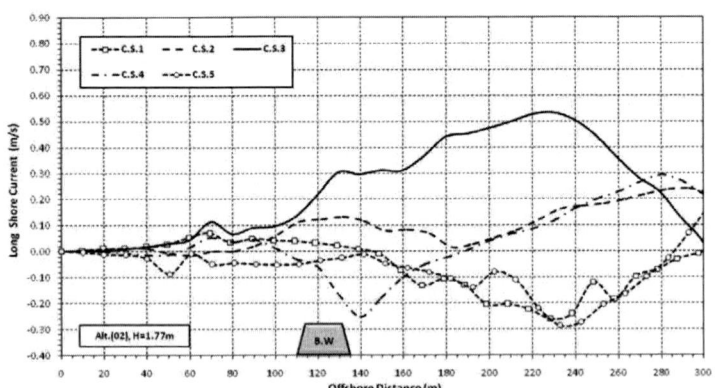

Figure A-10-a: Long shore current velocity along the transects (H=1.77m).

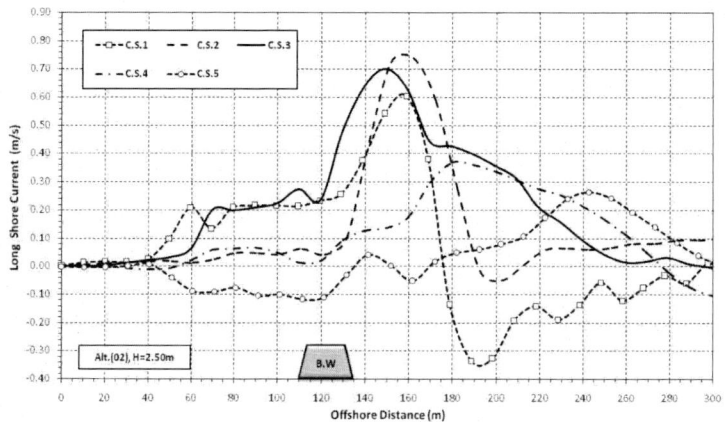

Figure A-10-b:Long shore current velocity along the transects (H=2.50m).

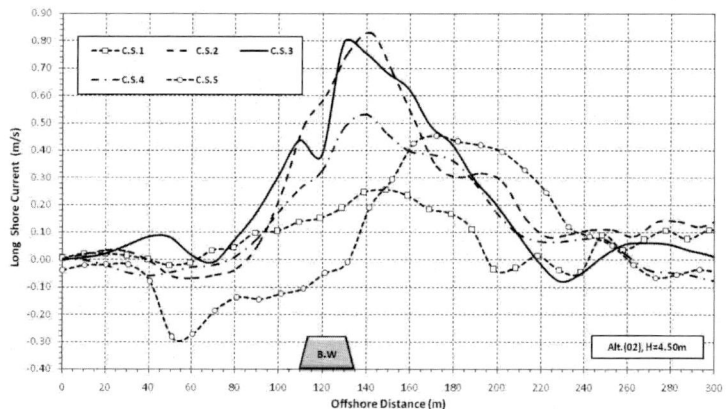

Figure A-10-c: Long shore current velocity along the transects (H=4.50m).

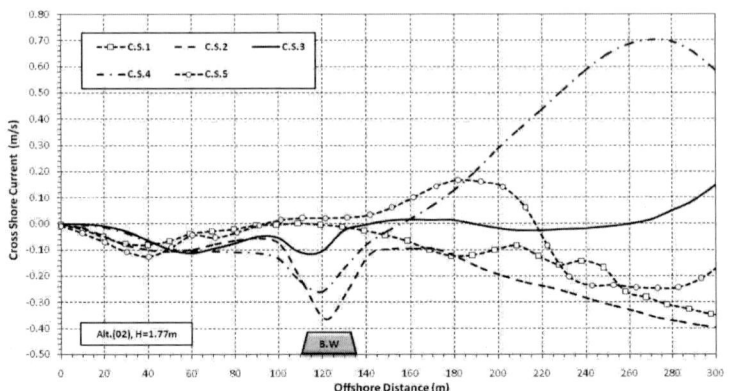

Figure A-11-a:Cross shore (rip) current velocity along the transects (H=1.77m).

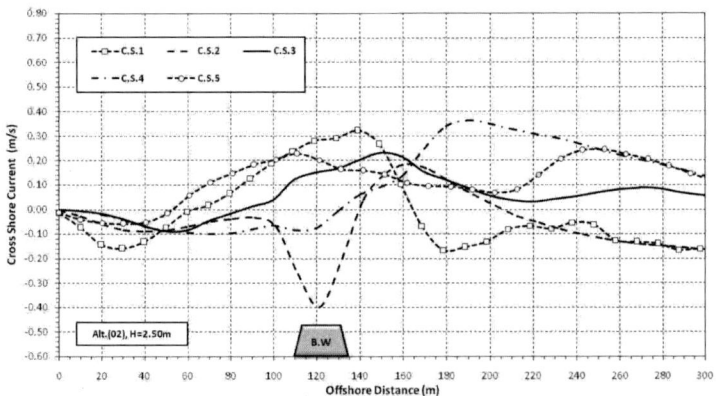

Figure A-11-b: Cross shore (rip) current velocity along the transects (H=2.50m).

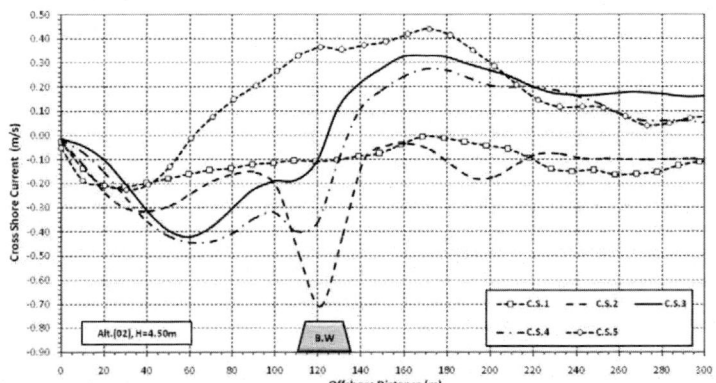

Figure A-12-c:Cross shore (rip) current velocity along the transects (H=4.50m).

Results of Alternative (03)

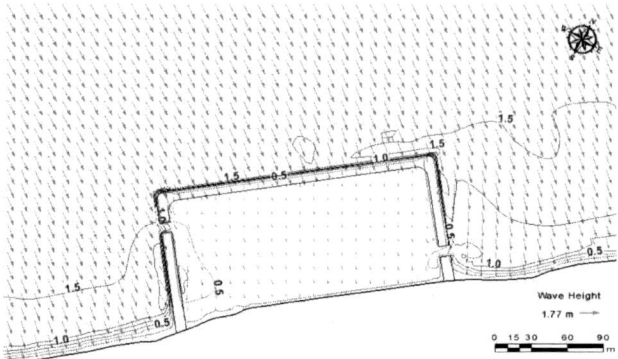

Figure A-13-a: Computed wave height and direction (H=1.77m).

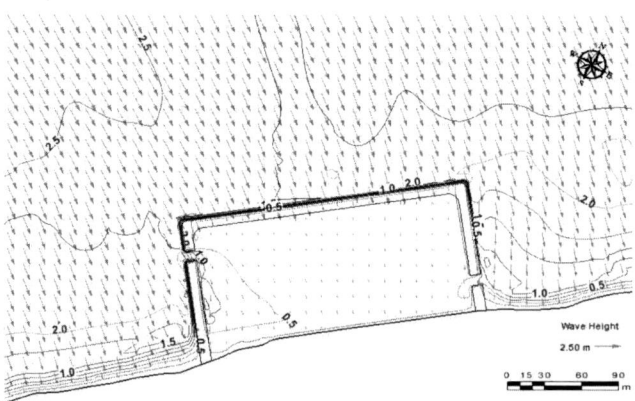

Figure A-13-b: Computed wave height and direction (H=2.50m).

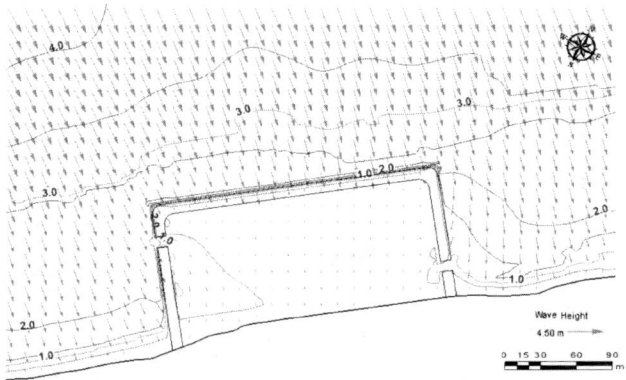

Figure A-13-c: Computed wave height and direction (H=4.50m).

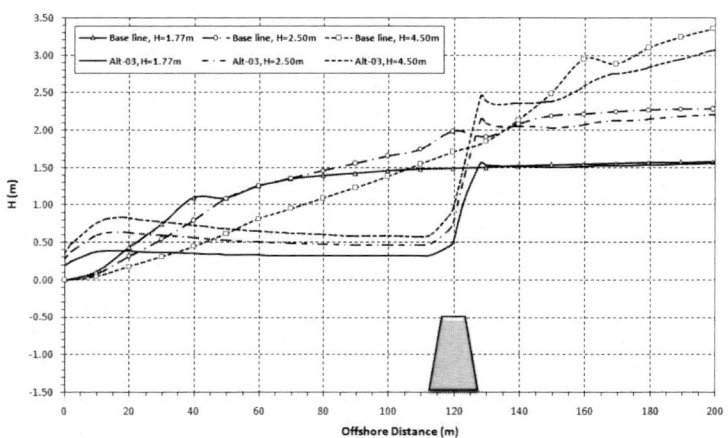

Figure A-14:Computed wave height along the centerline of the perched beach (at C.S.3).

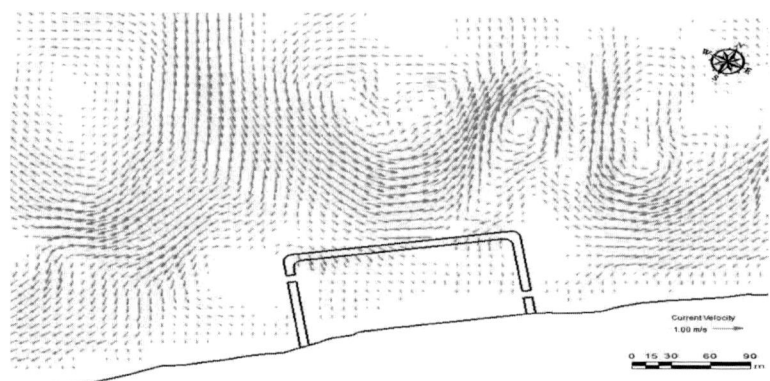

Figure A-15-a:Current velocity in the vicinity of the perched beach (H=1.77m).

Figure A-15-b:Current velocity in the vicinity of the perched beach (H=2.50m).

Figure A-15-c:Current velocity in the vicinity of the perched beach (H=4.50m).

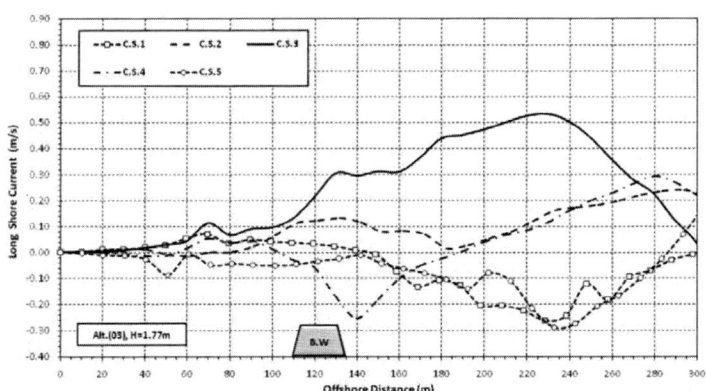

Figure A-16-a:Long shore current velocity along the transects (H=1.77m).

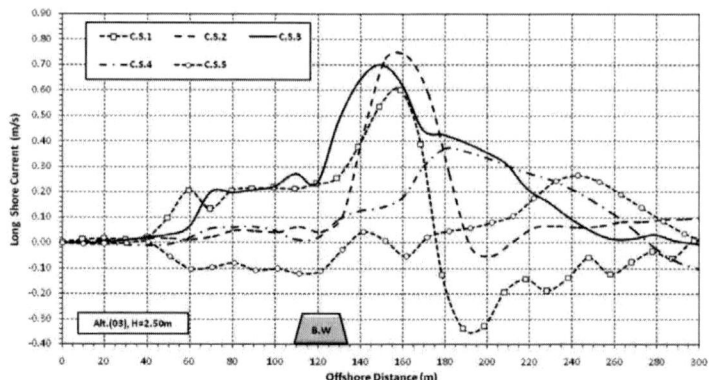

Figure A-16-b:Long shore current velocity along the transects (H=2.50m).

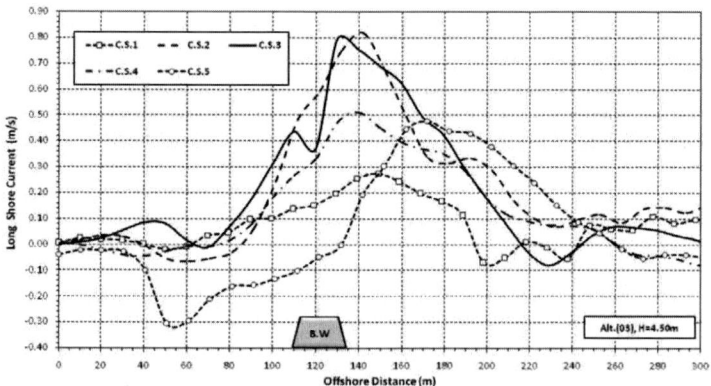

Figure A-16-c:Long shore current velocity along the transects (H=4.50m).

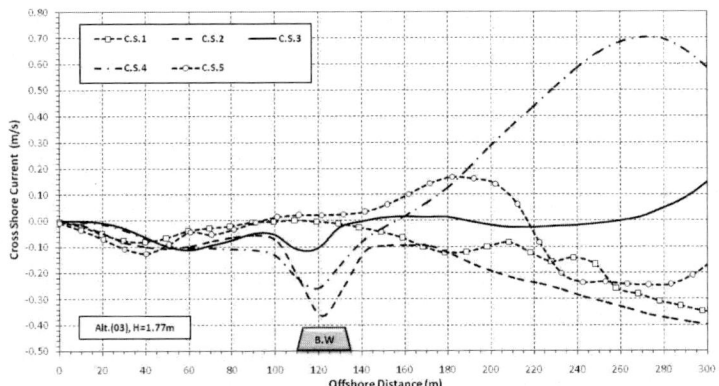

Figure A-17-a:Cross shore (rip) current velocity along the transects (H=1.77m).

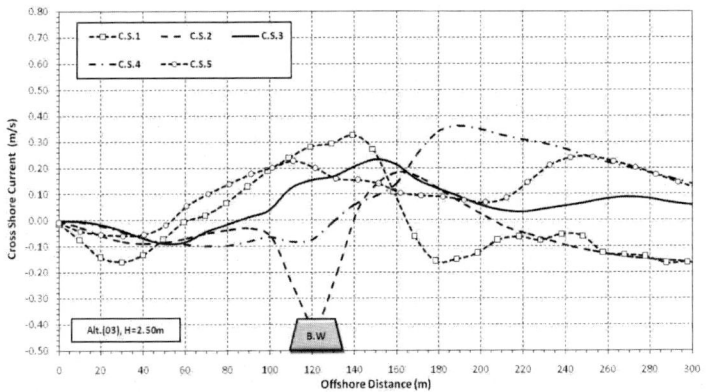

Figure A-17-b:Cross shore (rip) current velocity along the transects (H=2.50m).

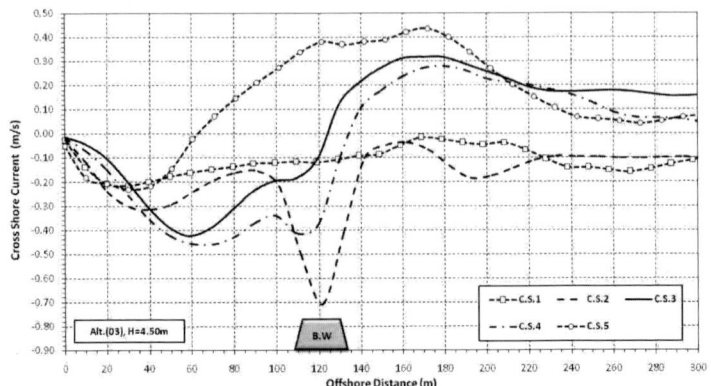

Figure A-17-c: Cross shore (rip) current velocity along the transects (H=4.50m).

Results of Alternative (04)

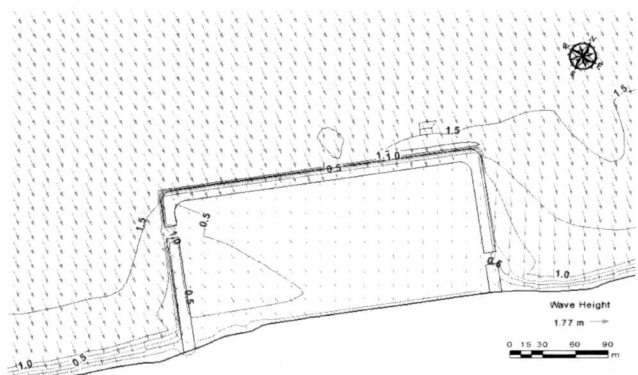

Figure A-18-a: Computed wave height and direction (H=1.77)m.

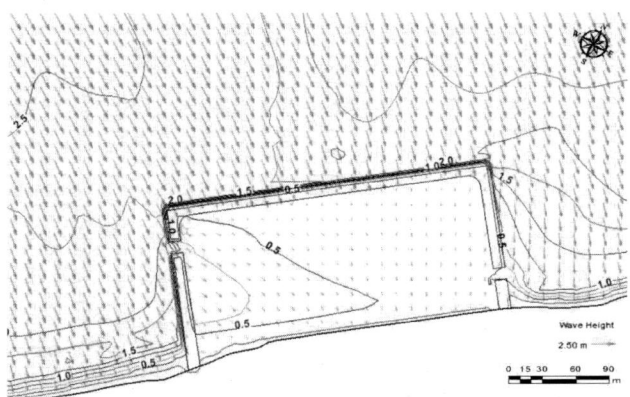

Figure A-18-b: Computed wave height and direction (H=2.50m).

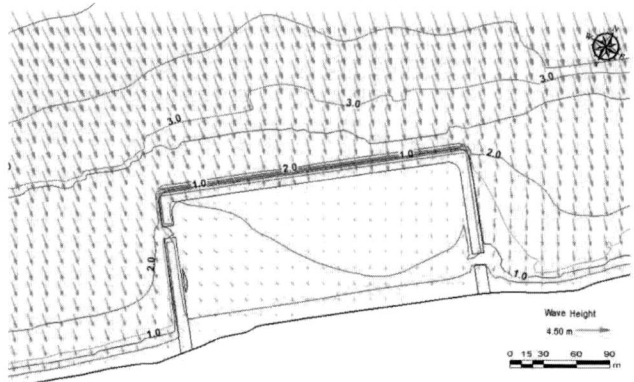

Figure A-18-c: Computed wave height and direction(H=4.50m).

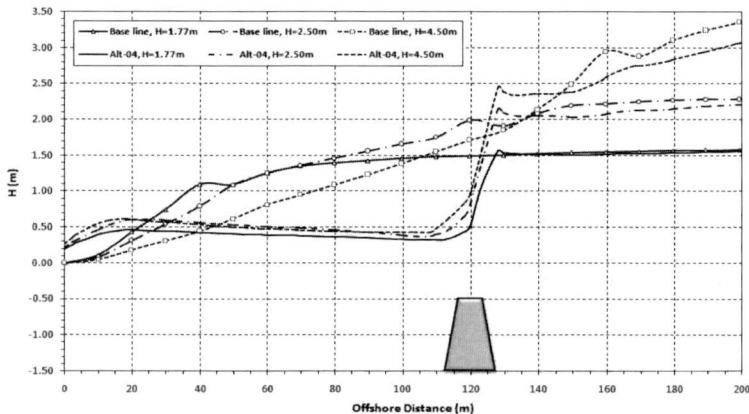

Figure A-19:Computed wave height along the centerline of the perched beach (at C.S.3).

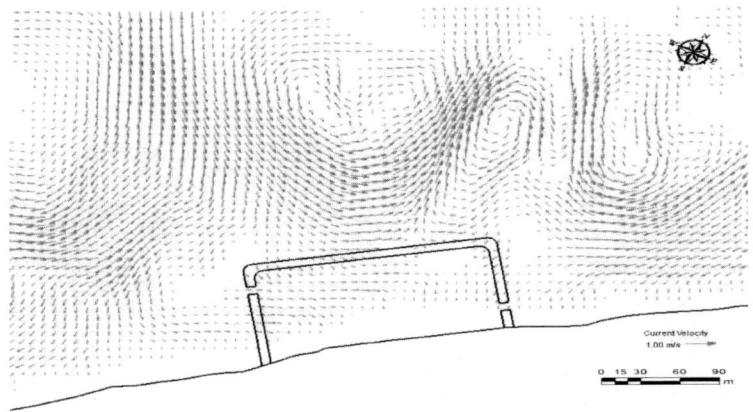

Figure A-20-a:Current velocity in the vicinity of the perched beach (H=1.77m).

Figure A-20-b:Current velocity in the vicinity of the perched beach (H=2.50m).

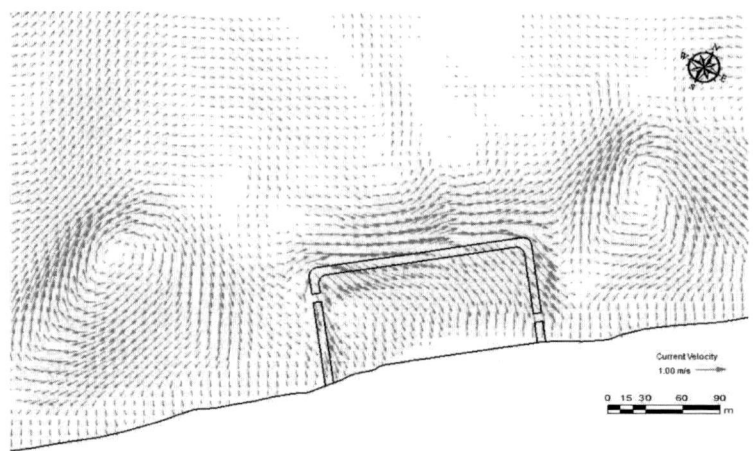

Figure A-20-c:Current velocity in the vicinity of the perched beach (H=4.50m).

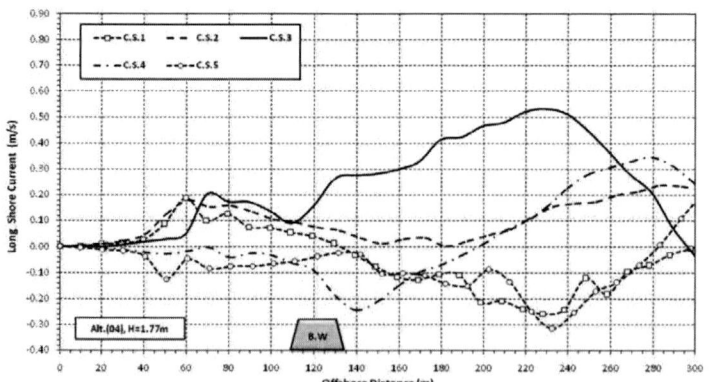

Figure A-21-a:Long shore current velocity along the transects (H=1.77m).

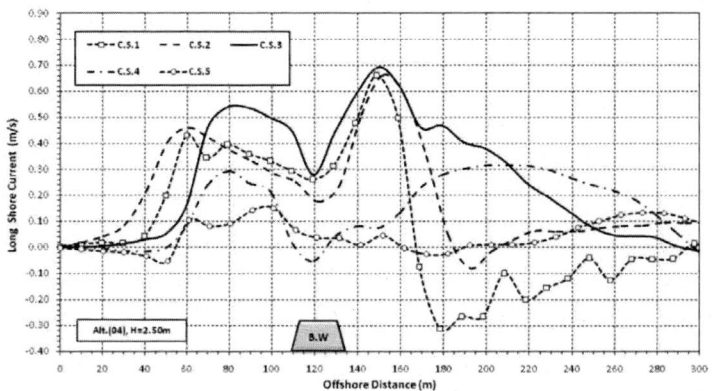

Figure A-21-b:Long shore current velocity along the transects (H=2.50m).

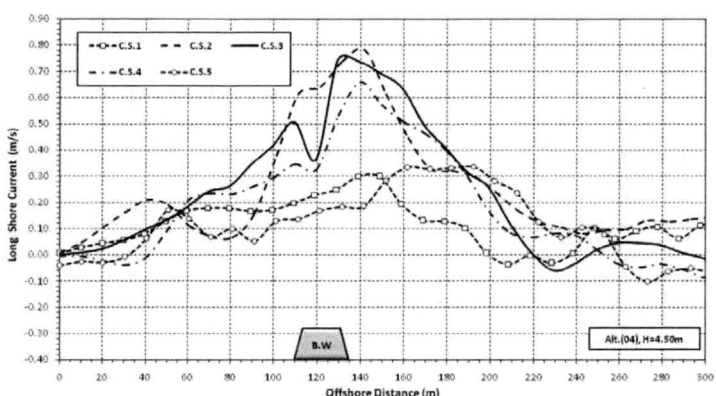

Figure A-21-c:Long shore current velocity along the transects (H=4.50m).

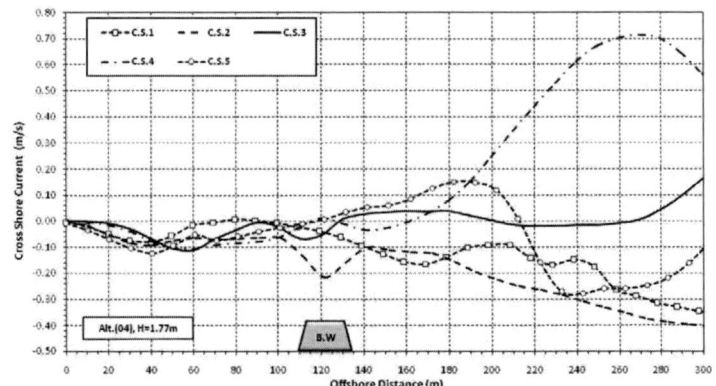

Figure A-22-a:Cross shore (rip) current velocity along the transects (H=1.77m).

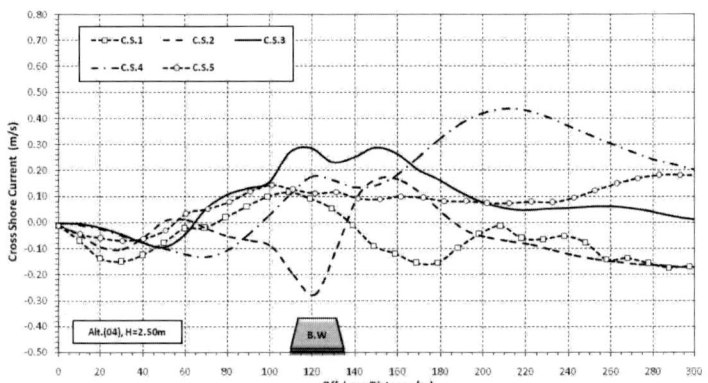

Figure A-22-b:Cross shore (rip) current velocity along the transects (H=2.50m).

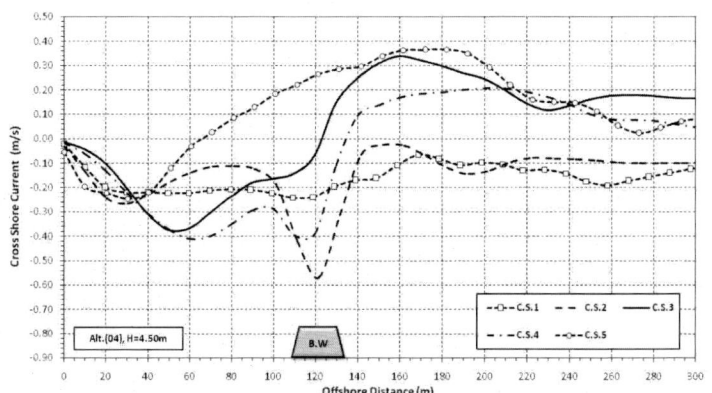

Figure A-22-c:Cross shore (rip) current velocity along the transects (H=4.50m).

Results of Alternative (05)

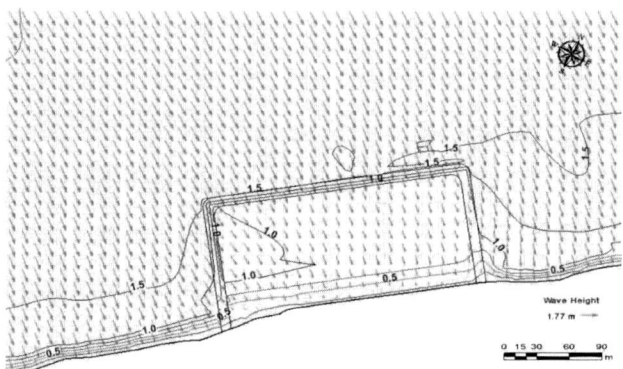

Figure A-23-a: Computed wave height and direction (H=1.77)m.

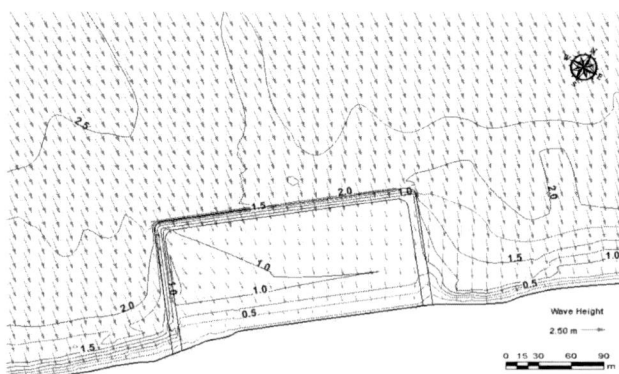

Figure A-23-b: Computed wave height and direction (H=2.50m).

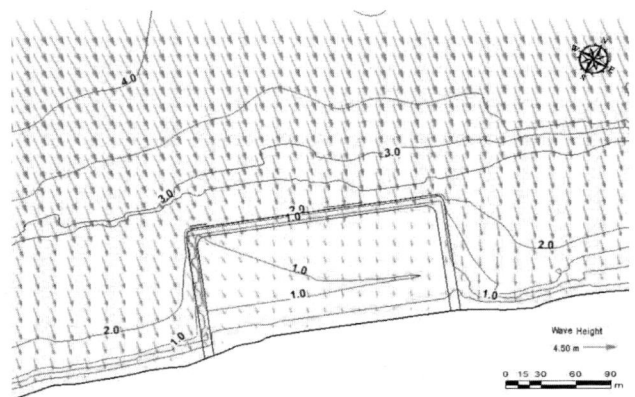

Figure A-23-c: Computed wave height and direction(H=4.50m).

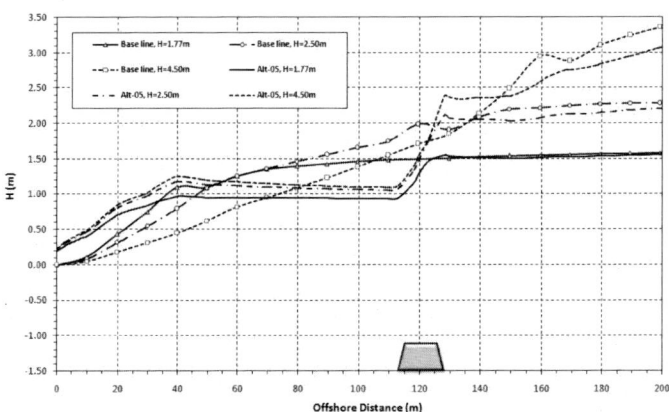

Figure A-24:Computed wave height along the centerline of the perched beach (at C.S.3).

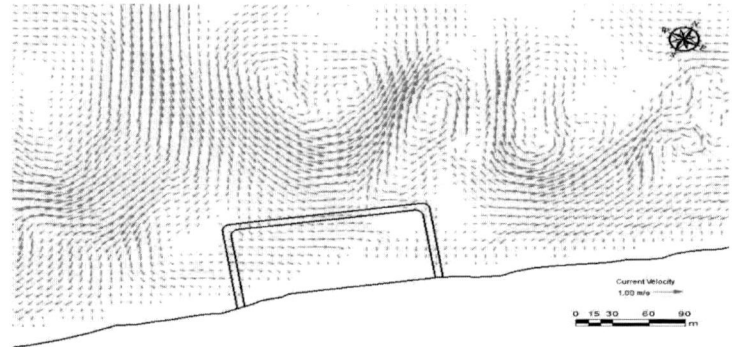

Figure A-25-a: Current velocity in the vicinity of the perched beach (H=1.77m).

Figure A-25-b: Current velocity in the vicinity of the perched beach (H=2.50m).

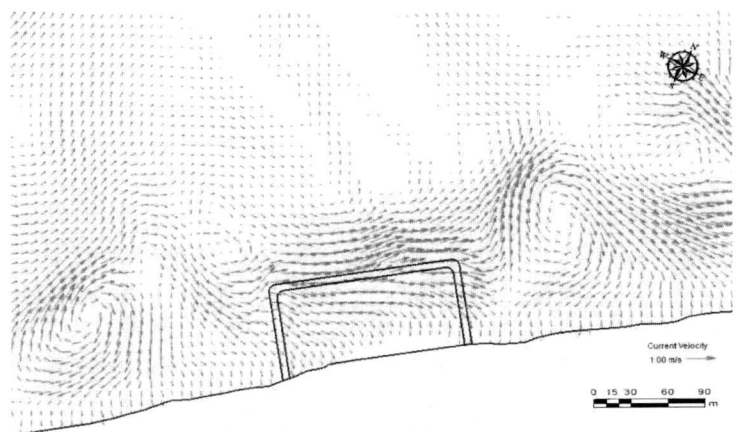

Figure A-25-c:Current velocity in the vicinity of the perched beach (H=4.50m).

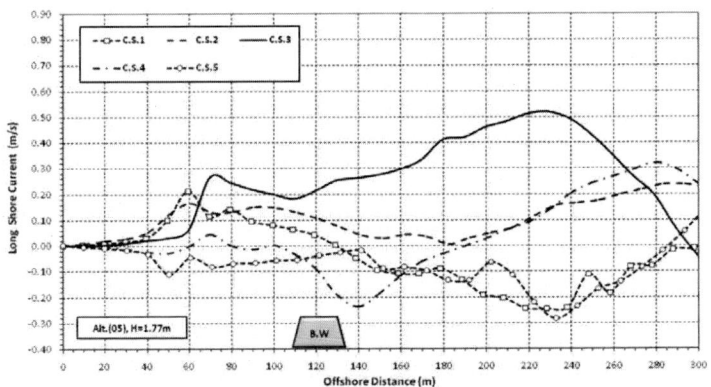

Figure A-26-a:Long shore current velocity along the transects (H=1.77m).

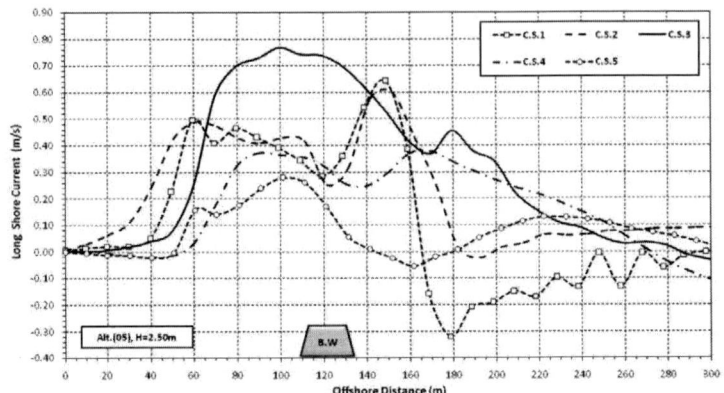

Figure A-26-b:Long shore current velocity along the transects (H=2.50m).

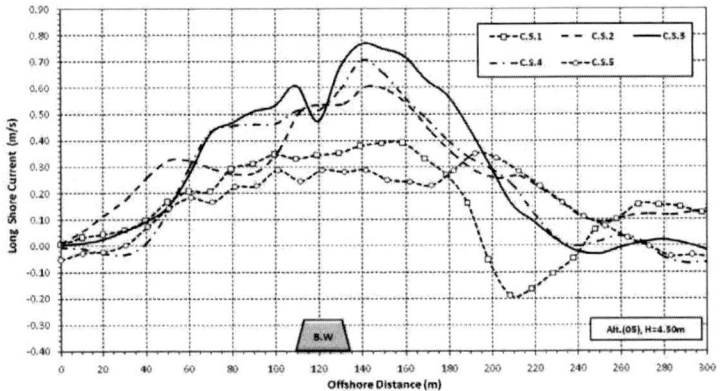

Figure A-26-c:Long shore current velocity along the transects (H=4.50m).

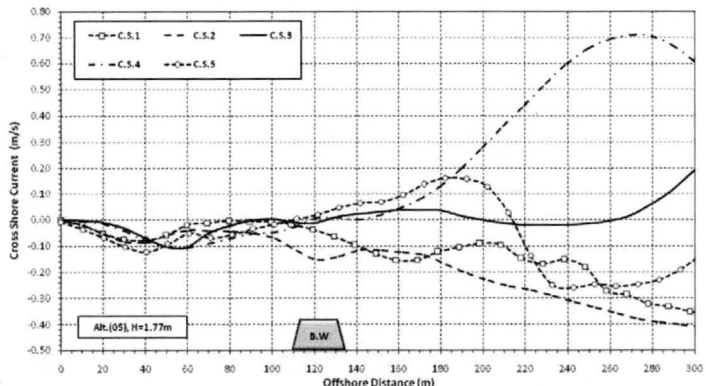

Figure A-27-a:Cross shore (rip) current velocity along the transects (H=1.77m).

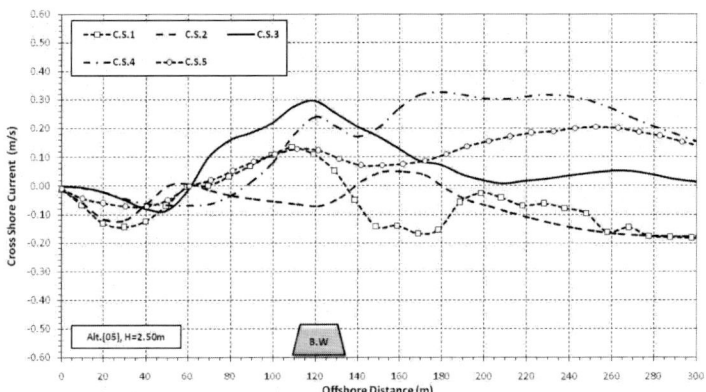

Figure A-27-b:Cross shore (rip) current velocity along the transects (H=2.50m).

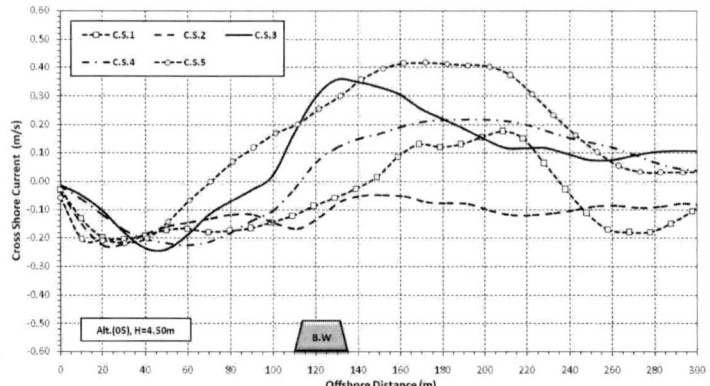

Figure A-27-c:Cross shore (rip) current velocity along the transects (H=4.50m).

Results of Alternative (06)

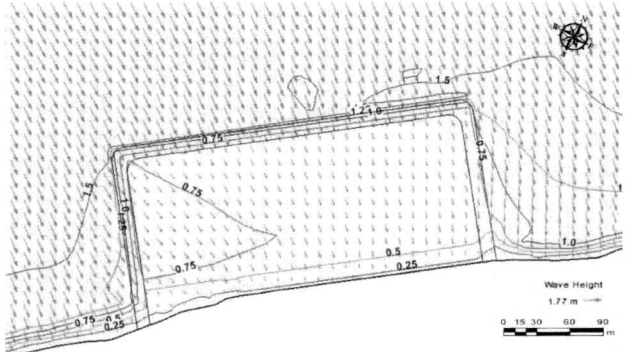

Figure A-28-a: Computed wave height and direction (H=1.77)m.

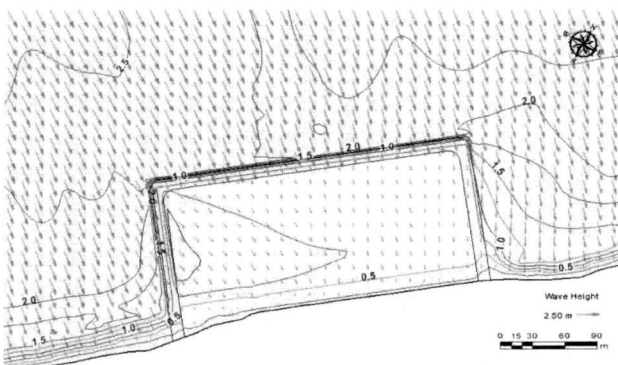

Figure A-28-b: Computed wave height and direction (H=2.50m).

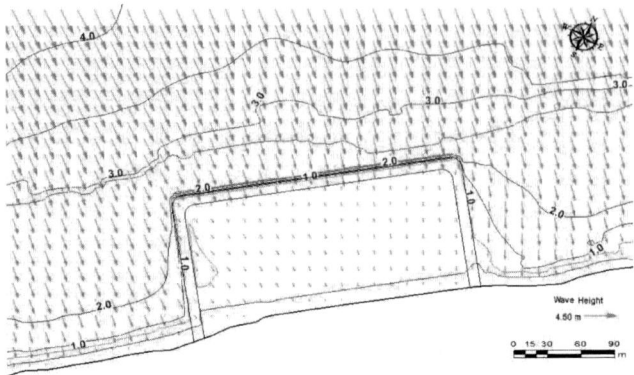

Figure A-28-c: Computed wave height and direction (H=4.50m).

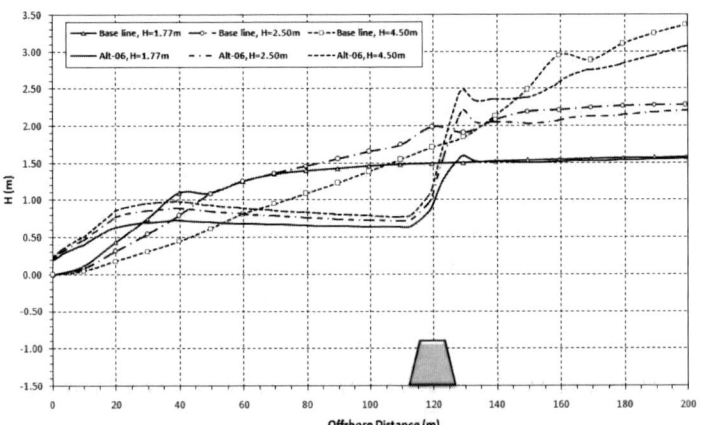

Figure A-29:Computed wave height along the centerline of the perched beach (at C.S.3).

Figure A-30-a:Current velocity in the vicinity of the perched beach (H=1.77m).

Figure A-30-b:Current velocity in the vicinity of the perched beach (H=2.50m).

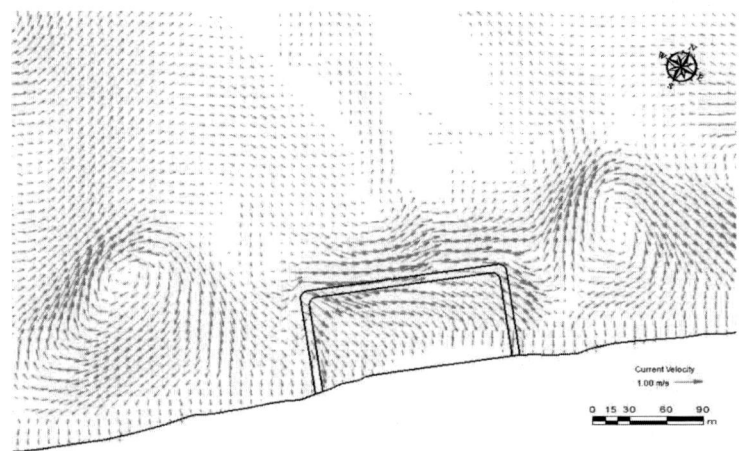

Figure A-30-c: Current velocity in the vicinity of the perched beach (H=4.50m).

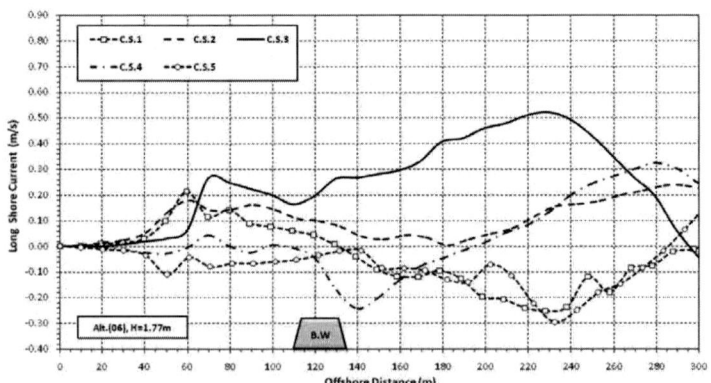

Figure A-31-a: Long shore current velocity along the transects (H=1.77m).

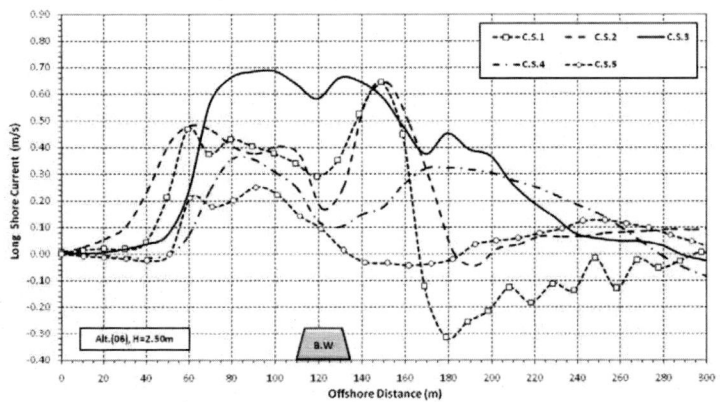

Figure A-31-b:Long shore current velocity along the transects (H=2.50m).

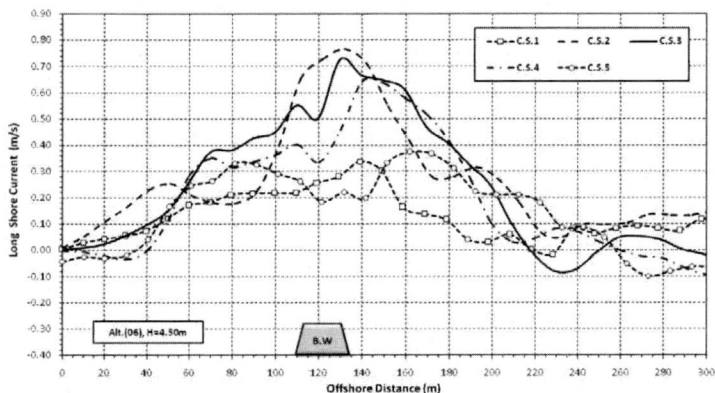

Figure A-31-c:Long shore current velocity along the transects (H=4.50m).

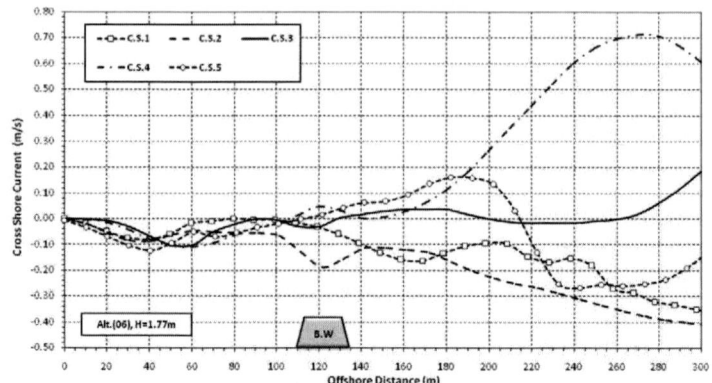

Figure A-32-a:Cross shore (rip) current velocity along the transects(H=1.77m).

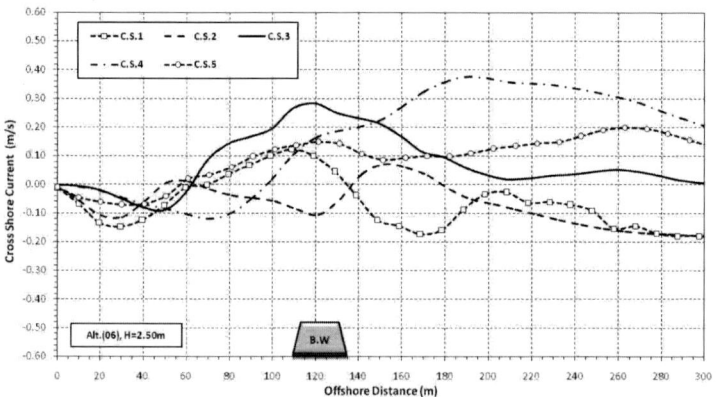

Figure A-32-b:Cross shore (rip) current velocity along the transects (H=2.50m).

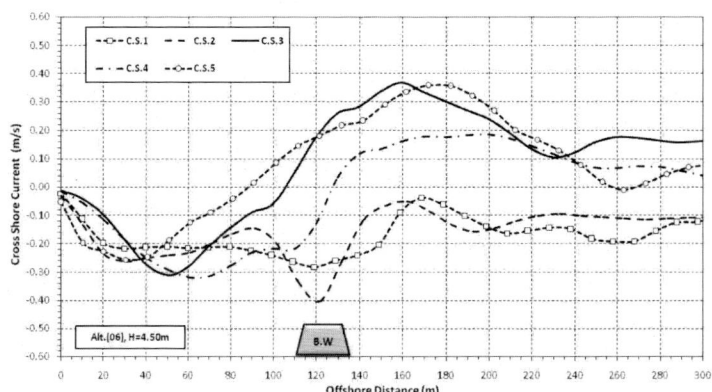

Figure A-32-c: Cross shore (rip) current velocity along the transects (H=4.50m).

Alternatives Comparison

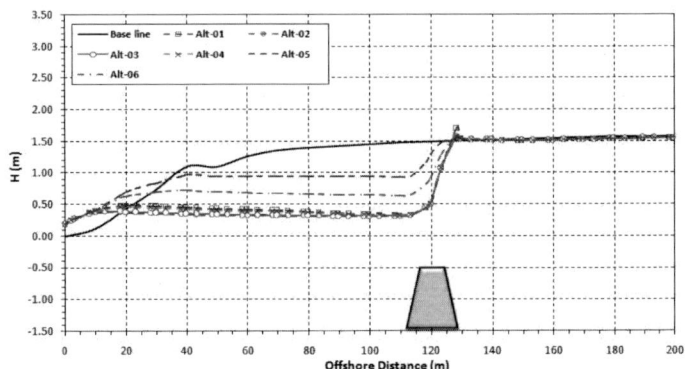

Figure A-33-a: Computed wave height along the centerline of the perched beach (H= 1.77m).

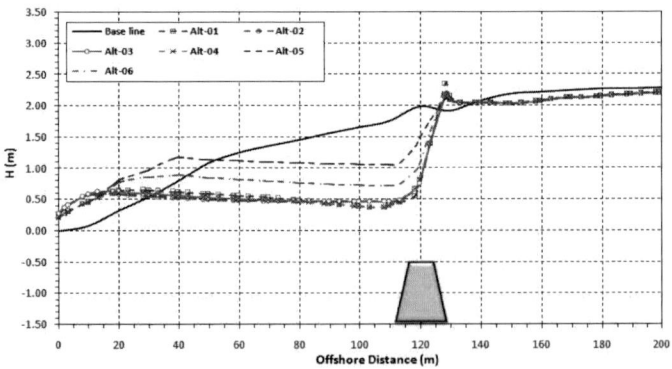

Figure A-33-b: Computed wave height along the centerline of the perched beach (H=2.50m).

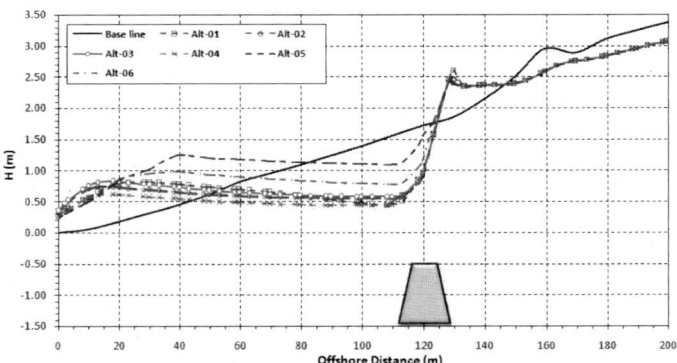

Figure A-33-c:Computed wave height along the centerline of the perched beach (H=4.50m).

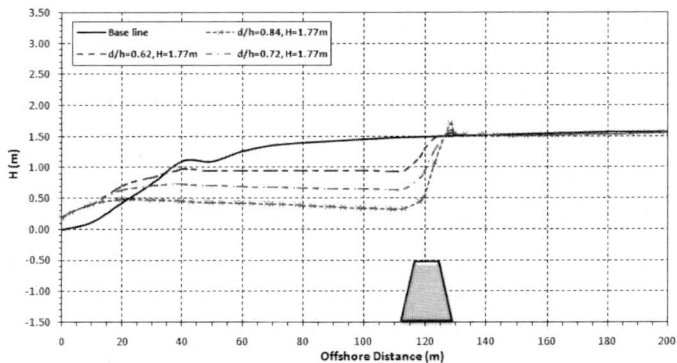

Figure A-34-a:Effect of d/h on the wave height at the centerline of the perched beach (H=1.77m).

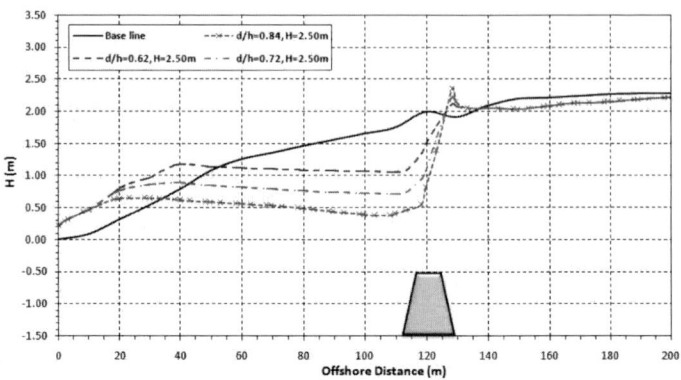

Figure A-34-b: Effect of d/h on the wave height at the centerline of the perched beach (H=2.50m).

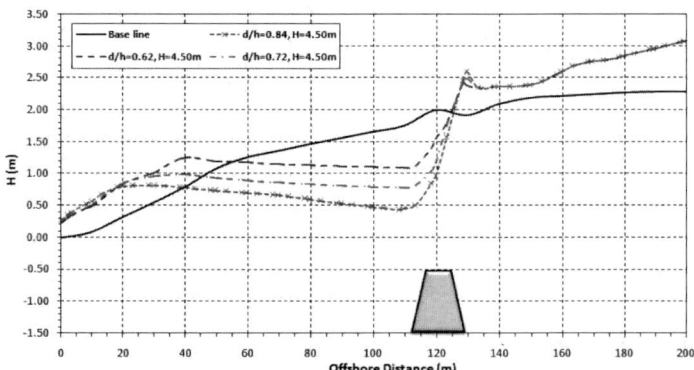

Figure A-34-c: Effect of d/h on the wave height at the centerline of the perched beach (H=4.50m).

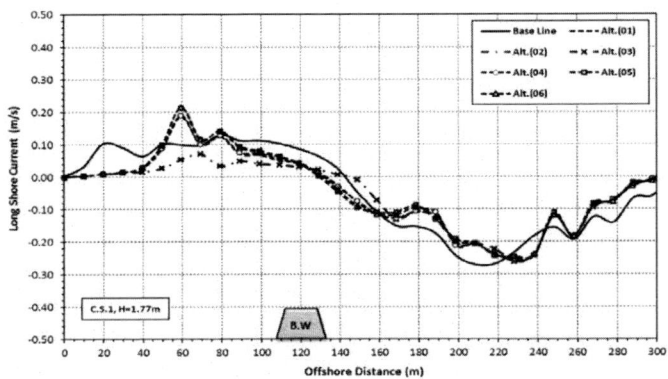

Figure A-35-a:Long shore current velocity along C.S.1(H= 1.77m).

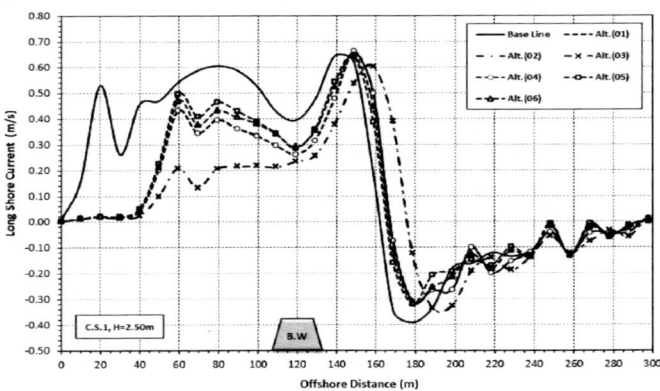

Figure A-35-b:Long shore current velocity along C.S.1 (H= 2.50m).

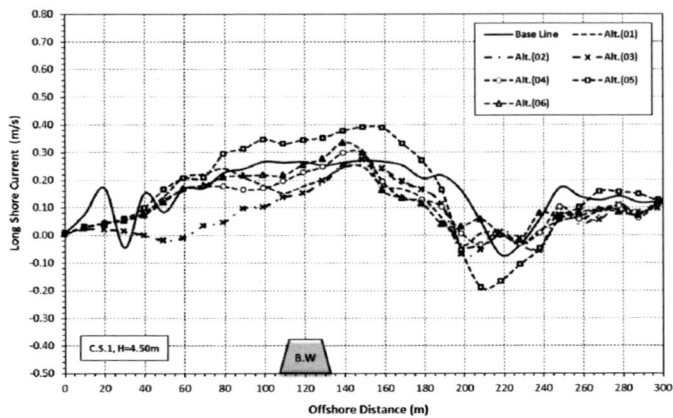

Figure A-35-c:Long shore current velocity along C.S.1 (H= 4.50m).

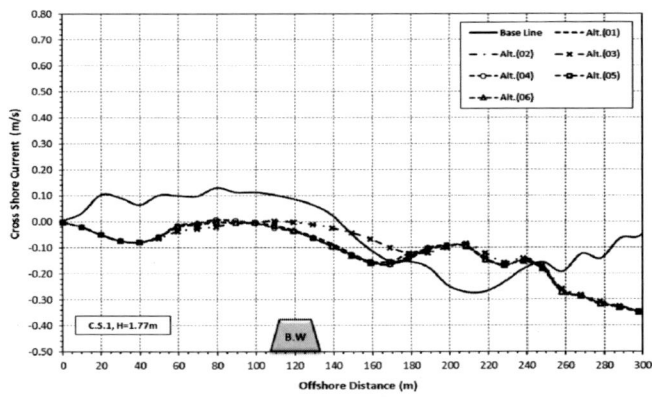

Figure A-35-d:Cross shore current velocity along C.S.1 (H= 1.77m).

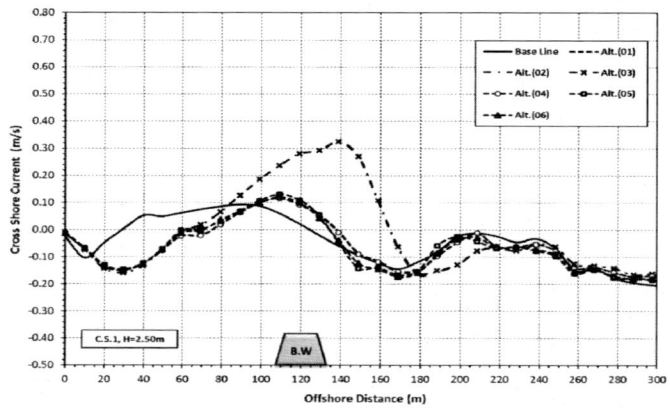

Figure A-35-e:Cross shore current velocity along C.S.1(H= 2.50m).

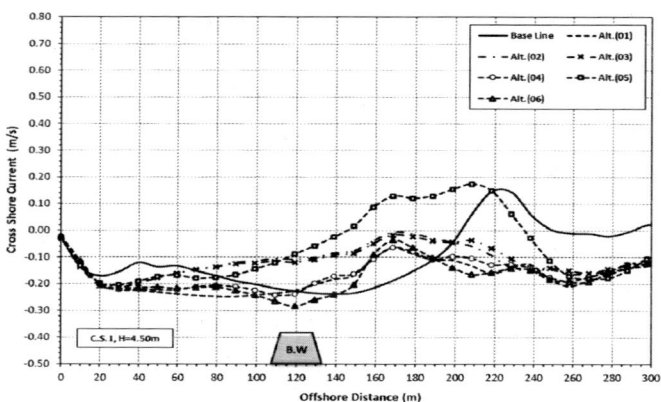

Figure A-35-f:Cross shore current velocity along C.S.1(H= 4.50m).

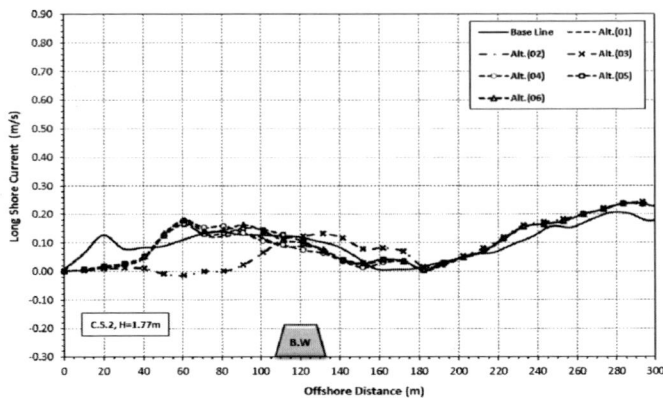

Figure A-36-a:Long shore current velocity along C.S.2 (H=1.77m).

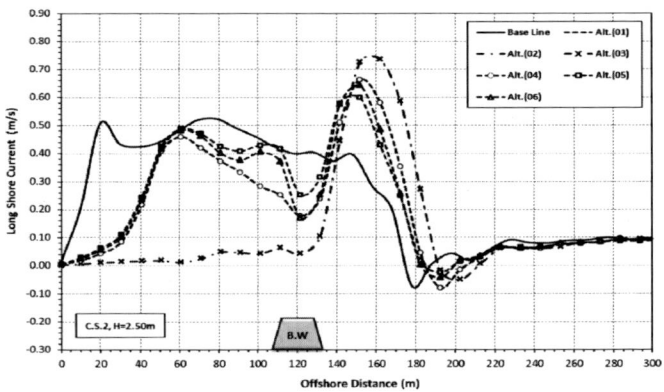

Figure A-36-b:Long shore current velocity along C.S.2 (H= 2.50m).

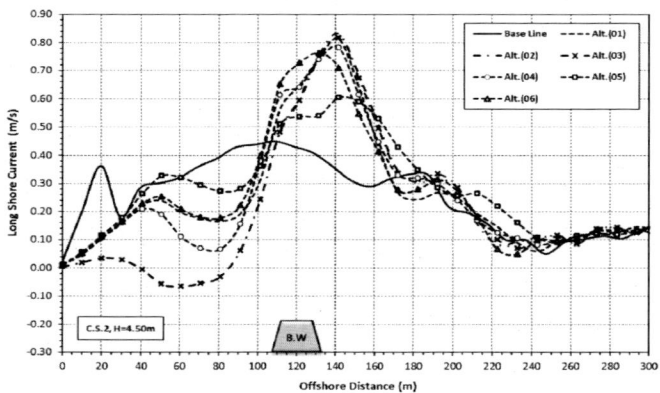

Figure A-36-c:Long shore current velocity along C.S.2(H= 4.50m).

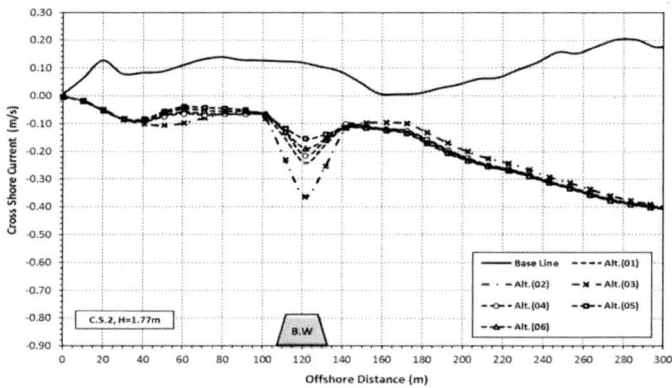

Figure A-36-d:Cross shore current velocity along C.S.2 (H= 1.77m).

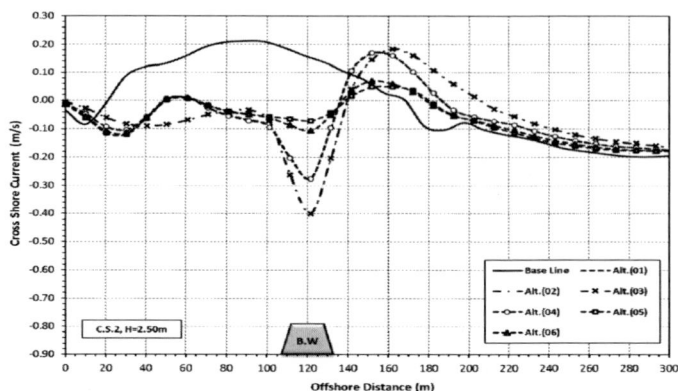

Figure A-36-e:Cross shore current velocity along C.S.2 (H= 2.50m).

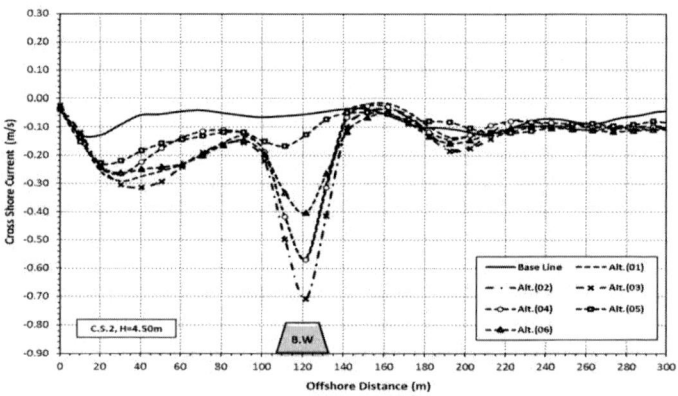

Figure A-36-f:Cross shore current velocity along C.S.2 (H= 4.50m).

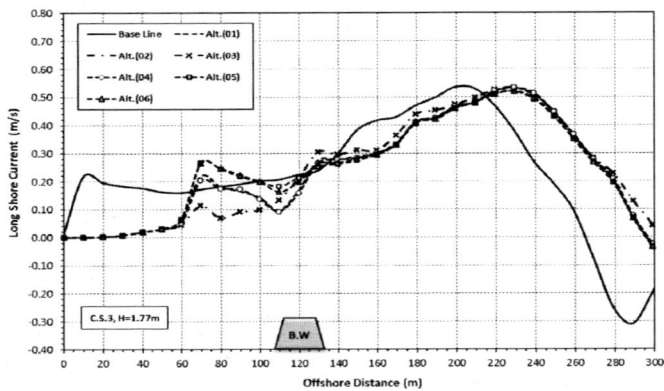

Figure A-37-a:Long shore current velocity along C.S.3 (H=1.77m).

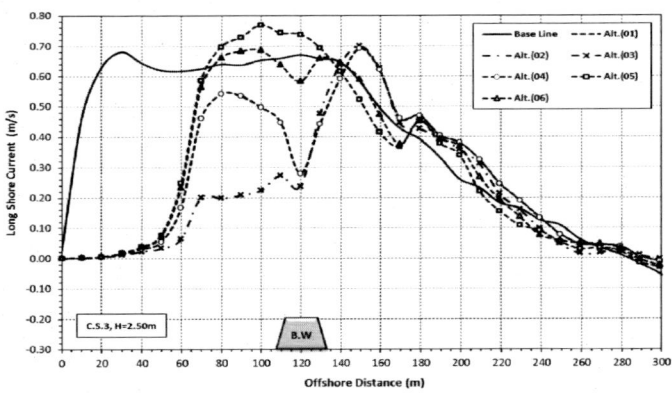

Figure A-37-b:Long shore current velocity along C.S.3 (H=2.50m).

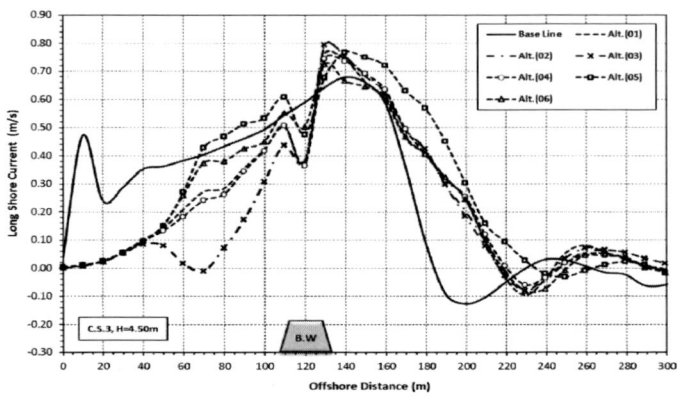

Figure A-37-c:Long shore current velocity along C.S.3 (H=4.50m).

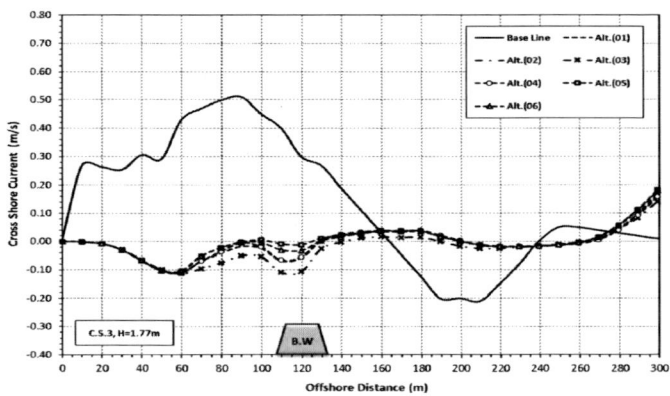

Figure A-37-d:Cross shore current velocity along C.S.3 (H= 1.77m).

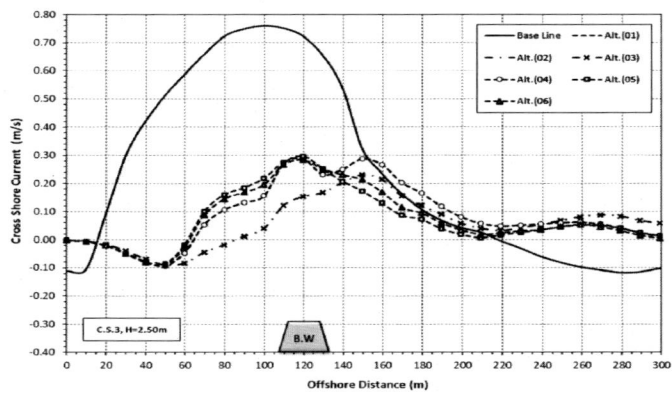

Figure A-37-e:Cross shore current velocity along C.S.3 (H= 2.50m).

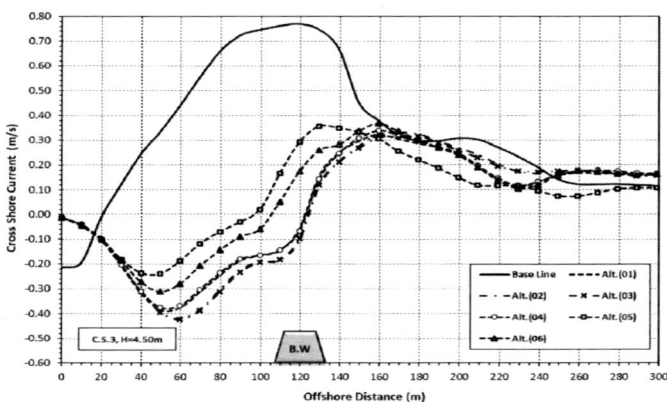

Figure A-37-f:Cross shore current velocity along C.S.3 (H= 4.50m).

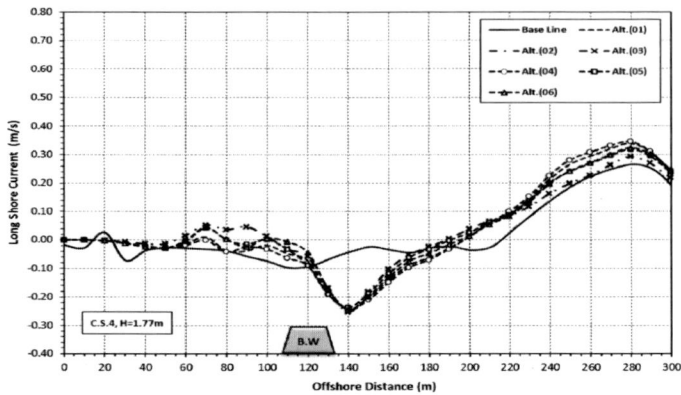

Figure A-38-a:Long shore current velocity along C.S.4 (H=1.77m).

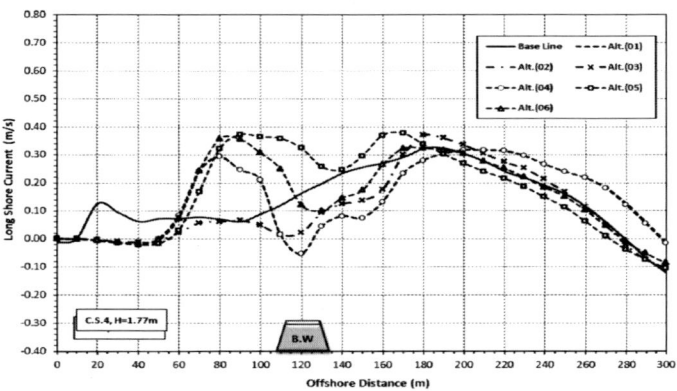

Figure A-38-b:Long shore current velocity along C.S.4 (H=2.50m).

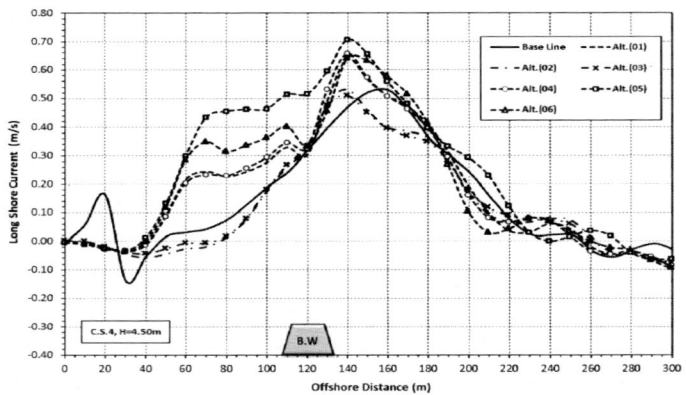

Figure A-38-c:Long shore current velocity along C.S.4 (H=4.50m).

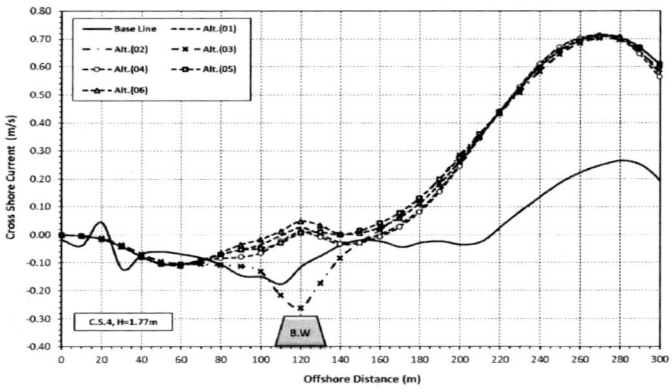

Figure A-38-d:Cross shore current velocity along C.S.4 (H= 1.77m).

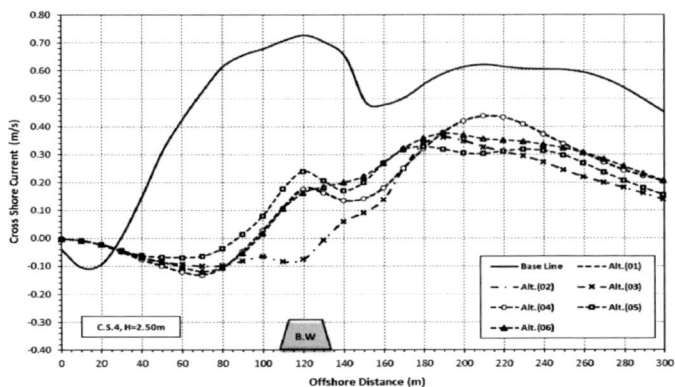

Figure A-38-e: Cross shore current velocity along C.S.4 (H= 2.50m).

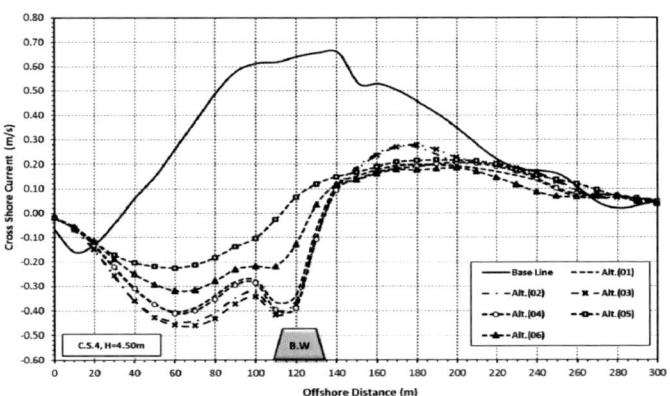

Figure A-38-f: Cross shore current velocity along C.S.4 (H= 4.50m).

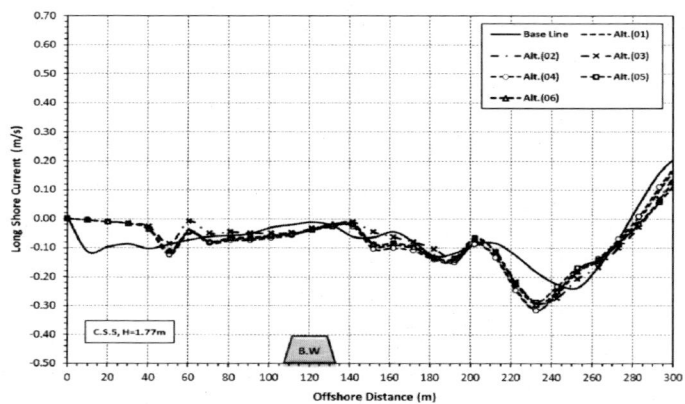

Figure A-39-a:Long shore current velocity along C.S.5 (H=1.77m).

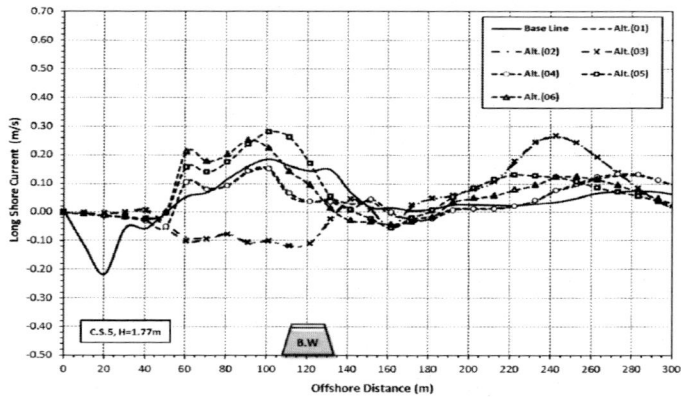

Figure A-39-b:Long shore current velocity along C.S.5 (H=2.50m).

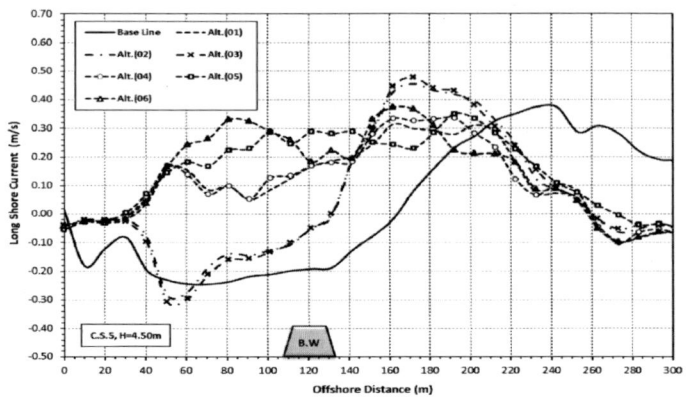

Figure A-39-c:Long shore current velocity along C.S.5 (H=4.50m).

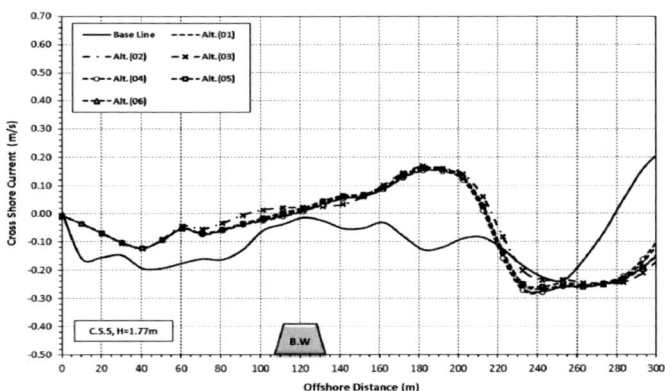

Figure A-39-d:Cross shore current velocity along C.S.5 (H= 1.77m).

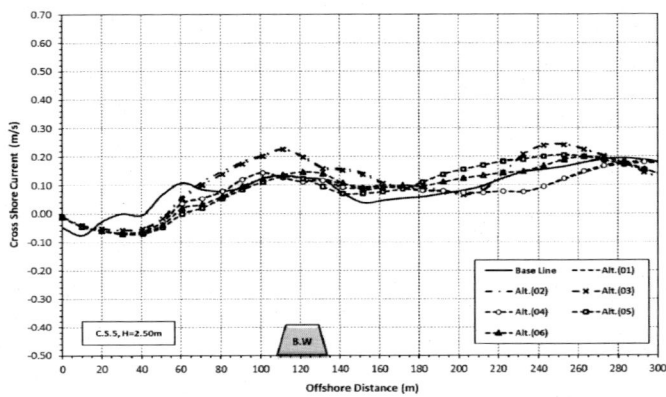

Figure A-39-e:Cross shore current velocity along C.S.5 (H= 2.50m).

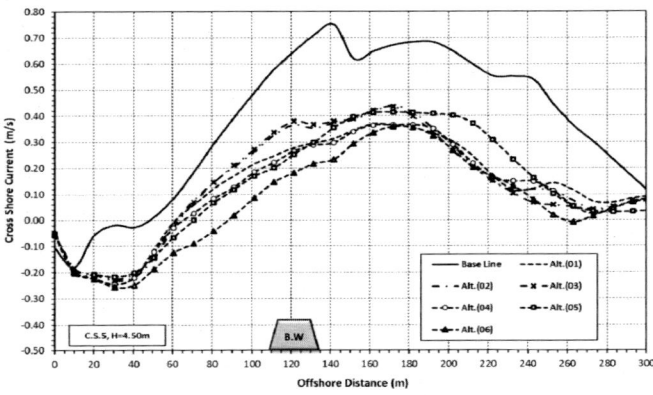

Figure A-39-f: Cross shore current velocity along C.S.5 (H= 4.50m).

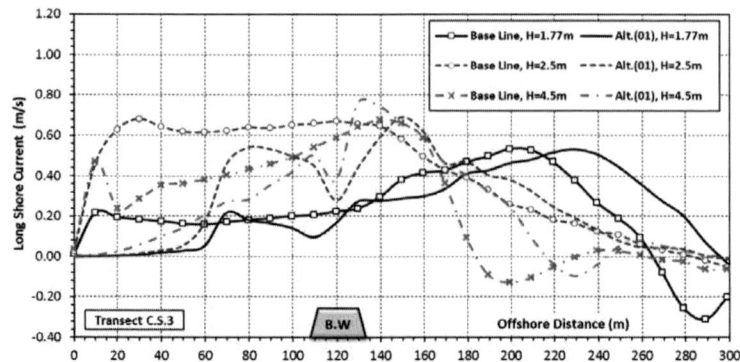

Figure A-40-a: Long shore current velocity along C.S.3 before/after construction of the perched beach (Alt. 01).

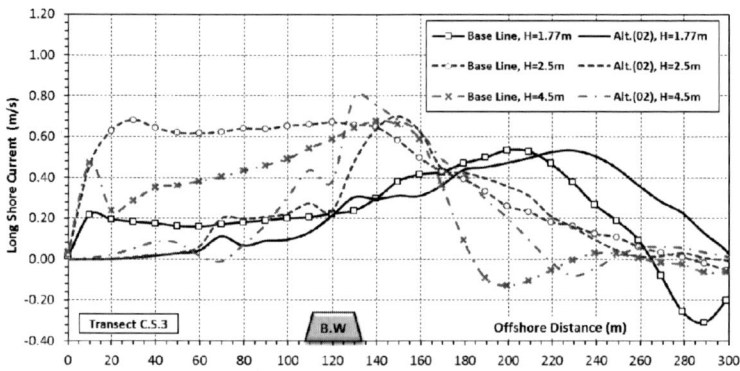

Figure A-40-b: Long shore current velocity along C.S.3 before/after construction of the perched beach (Alt. 02).

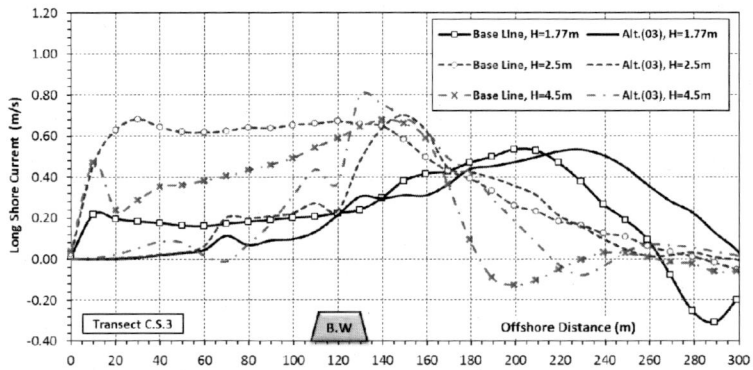

Figure A-40-c: Long shore current velocity along C.S.3 before/after construction of the perched beach (Alt. 03).

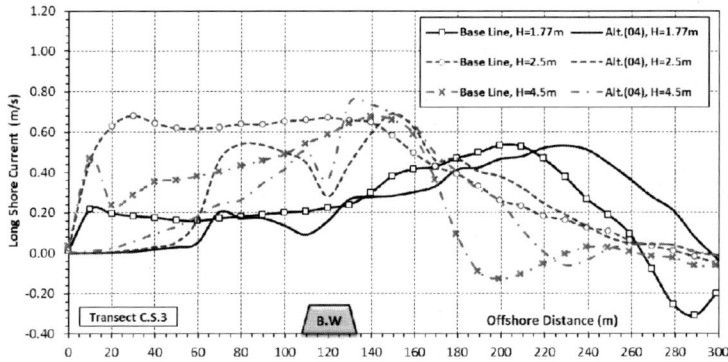

Figure A-40-d: Long shore current velocity along C.S.3 before/after construction of the perched beach (Alt. 04).

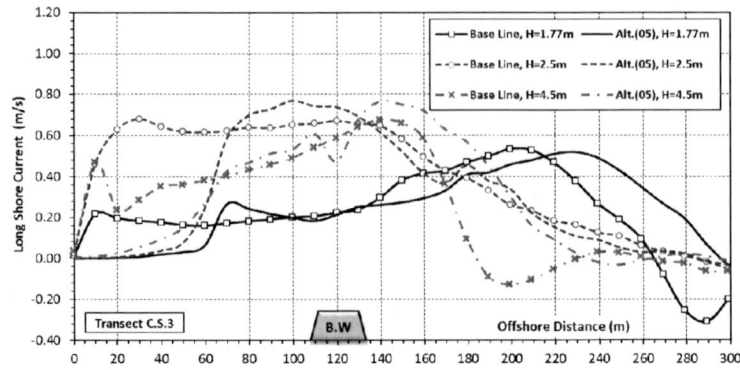

Figure A-40-e: Long shore current velocity along C.S.3 before/after construction of the perched beach (Alt. 05).

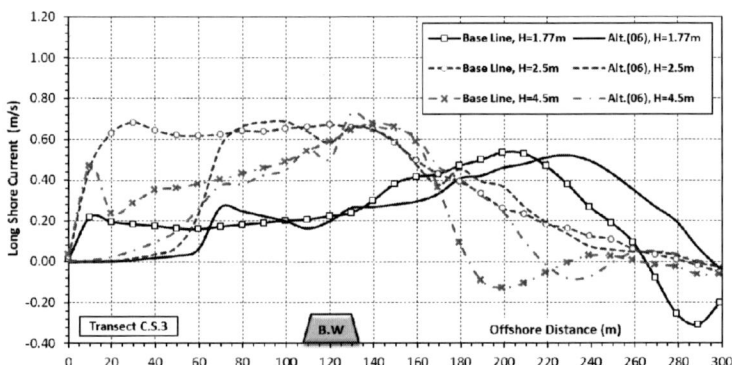

Figure A-40-f: Long shore current velocity along C.S.3 before/after construction of the perched beach (Alt. 06).

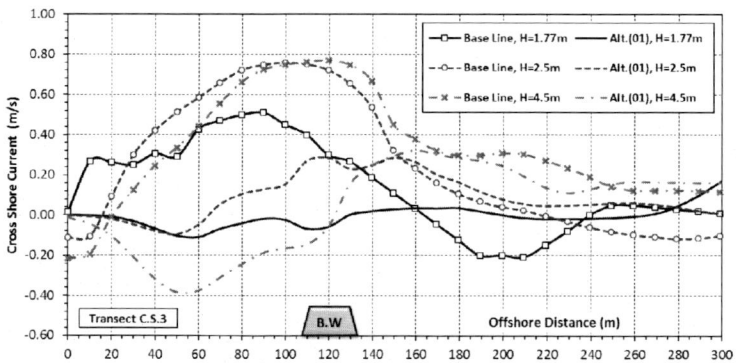

Figure A-41-a: Rip current velocity along C.S.3 before/after construction of the perched beach (Alt. 01).

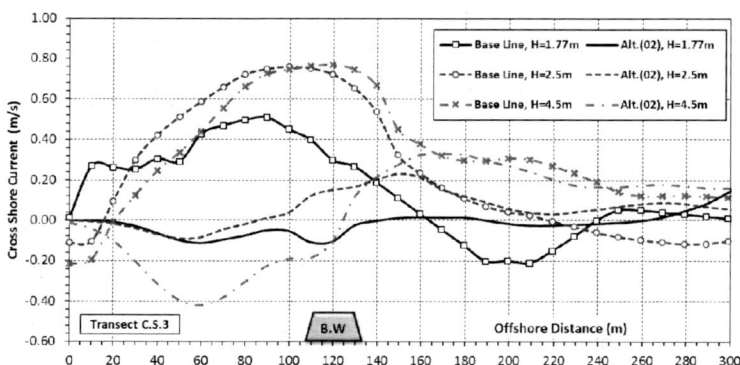

Figure A-41-b: Rip current velocity along C.S.3 before/after construction of the perched beach (Alt. 02).

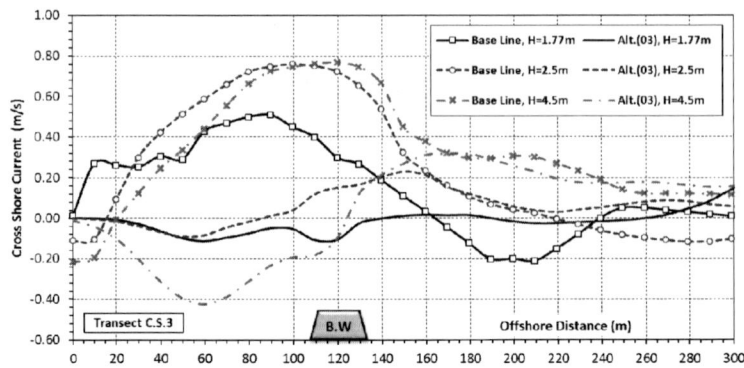

Figure A-41-c: Rip current velocity along C.S.3 before/after construction of the perched beach (Alt. 03).

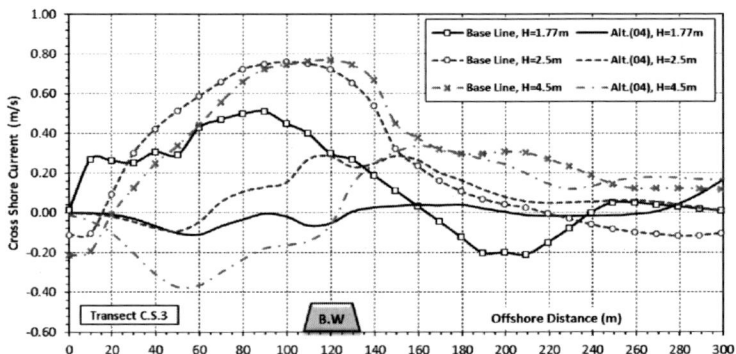

Figure A-41-d: Rip current velocity along C.S.3 before/after construction of the perched beach (Alt. 04).

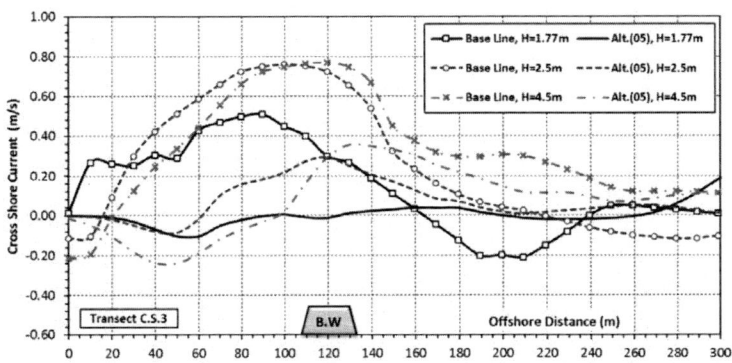

Figure A-41-e: Rip current velocity along C.S.3 before/after construction of the perched beach (Alt. 05).

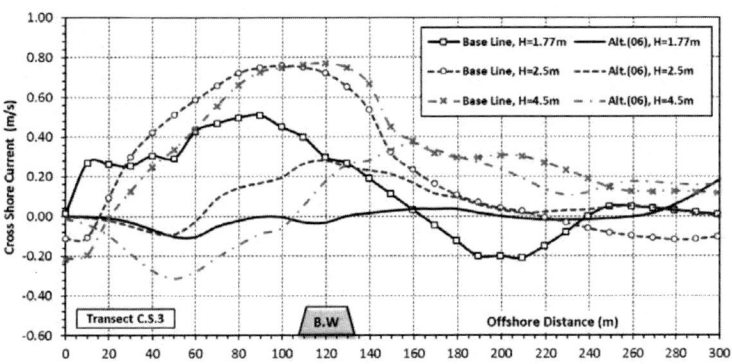

Figure A-41-f: Rip current velocity along C.S.3 before/after construction of the perched beach (Alt. 06).

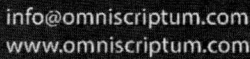